THE
MATHEMATICS
DEVOTIONAL

CELEBRATING *the* WISDOM *and* BEAUTY *of* MATHEMATICS

CLIFFORD A. PICKOVER

STERLING
New York

STERLING
New York

An Imprint of Sterling Publishing
387 Park Avenue South
New York, NY 10016

ISBN 978-1-4549-1322-1

Distributed in Canada by Sterling Publishing
℅ Canadian Manda Group, 165 Dufferin Street
Toronto, Ontario, Canada M6K 3H6
Distributed in the United Kingdom by GMC Distribution Services
Castle Place, 166 High Street, Lewes, East Sussex, England BN7 1XU
Distributed in Australia by Capricorn Link (Australia) Pty. Ltd.
P.O. Box 704, Windsor, NSW 2756, Australia

For information about custom editions, special sales, and premium and corporate
purchases, please contact Sterling Special Sales at 800-805-5489
or specialsales@sterlingpublishing.com.

Manufactured in China

2 4 6 8 10 9 7 5 3 1

www.sterlingpublishing.com

CONTENTS

∞

INTRODUCTION

A Universe of Patterns

Readers of my popular mathematics books already know how I feel about numbers and mathematics. Both are portals to other universes and new ways of thinking. In some sense, numbers help us glimpse a realm partly shielded from our minds, which have not evolved to fully comprehend the mathematical fabric. This tapestry stretches, in practical and theoretical areas, like a vast spiderweb with an infinity of connections and patterns. Higher mathematical discussions are a little like poetry. Danish physicist Niels Bohr felt the same way about physics when he said, "We must be clear that, when it comes to atoms, language can be used only as in poetry."

As evident in many of the quotations selected for this book, mathematicians, throughout history, have often approached mathematics with a sense of awe, reverence, and mystery. Some of the individuals who are quoted have likened mathematicians to artists who seek patterns and who have aesthetic perspectives on their subject matter. Throughout history, some mathematicians also had an almost mystical or religious sense about their work. For example, Indian mathematician Srinivasa Ramanujan (1887–1920) seemed to pluck mathematical ideas from the ether, out of his dreams. He thought the gods gave him his insights. He could read the codes in the mathematical matrix in the same way that Neo, the lead character in the movie *The Matrix*, could access mathematical symbols cascading about him that formed the infrastructure of reality. I don't know if God is a cryptographer, but the codes and patterns are all "around" us, waiting to be deciphered. Some may take a thousand years to understand. Some may always be shrouded in mystery.

As I point out in my book *A Passion for Mathematics*, many other mathematicians—like Carl Friedrich Gauss, James Hopwood Jeans, Georg Cantor, Blaise Pascal, and John Littlewood—believed that mathematical inspiration had a divine aspect. Gauss said he once proved a theorem "not by dint of painful effort but so to speak by the grace of God." For these reasons, I have included a number of quotations that touch on this mystical aspect—to dispel the notion that mathematics is totally separate from such disparate realms as religion, poetry, and art.

Today, computers with graphics can be used to produce visual representations of mathematical behavior from a number of perspectives, with the potential for providing both seasoned mathematicians and laypeople with a sense of the behavior of even simple formulas that nonetheless can exhibit complex behavior—behavior that was very difficult to fully appreciate before the age of the computer. Many of the images in this book represent fractals, an intricate-looking set of curves and shapes, most of which were never seen before the advent of computers, with their ability to quickly perform massive calculations. Fractals often exhibit self-similarity, which suggests that various exact or inexact copies of an object can be found in the original object at smaller sizes. The detail continues for many magnifications—like an endless nesting of Russian dolls within dolls. Some of these shapes exist only in abstract geometric space, but others can be used as models for complex natural objects, such as coastlines and blood vessel branching. Interestingly, fractals can provide a useful framework for understanding chaotic processes. The dazzling computer-generated images may be intoxicating, perhaps motivating students' interest in math more than any other mathematical discovery in the previous century. Physicists are interested in fractals because they can sometimes describe the chaotic behavior of real-world phenomena, such as planetary motion, fluid flow, the diffusion of drugs, the behavior of interindustry relationships, and the vibration of airplane wings. For additional diversity, other figures in this book relate to various forms of algorithmic art, knots, symmetries, tilings, optical illusions, tessellations, mazes, and natural forms that exhibit symmetrical patterns.

Of course, mathematical patterns rendered with computer graphics may be admired purely for their artistic appeal. The line between science, mathematics, and art can sometimes be a fuzzy one; the three are fraternal philosophies formalized, in part, by ancient Greeks like Pythagoras and Ictinus. Today, computer graphics is one method through which scientists, mathematicians, and artists reunite these disciplines by providing scientific and artistic ways to represent the behavior of mathematical functions. In 1987, Sven Carlson wrote in *Science News:*

> Art and science will eventually be seen to be as closely connected as arms to the body. Both are vital elements of order and its discovery. The word "art" derives from the Indo-European base "ar," meaning to join or fit together. In this sense, science, in the attempt to learn how and why things fit, becomes art. And when art is seen as the ability to do, make, apply or portray in a way that withstands the test of time, its connection with science becomes more clear.

Readers will note that the quotations in this book come from diverse sources, and although many are from important mathematicians, I have intentionally tried to include a range of sources—from serious mathematicians and educators to novelists like Stephen King and Leo Tolstoy. Even diverse thinkers such as Sir Winston Churchill and Emperor Napoléon Bonaparte make brief appearances.

This devotional is a way of providing glimpses of mathematics and aesthetics—presenting wisdom and poetry in brief, along with some mathematical "eye candy" in the form of computer graphics, hopefully inspiring readers to learn more about the universe of mathematics and the delights that mathematicians, artists, and computer programmers feel in exploring mathematics. I hope there are future editions of this book, and welcome suggestions and corrections from readers.

Of course, although this book is not a "daily devotional" in the traditional religious sense of the phrase, I hope that contemplating the quotations and images in this book fills your mind with wonder and astonishment—while stretching your imagination and serving as a source of inspiration and beauty. Going beyond inspiration, the *usefulness* of mathematics allows us to build spaceships and investigate the geometry of our universe. Numbers will be our first means of communication with intelligent alien races. Today, mathematics has permeated every field of scientific endeavor and plays an invaluable role in biology, physics, chemistry, economics, sociology, and engineering. Math can be used to help explain the structure of a rainbow, teach us how to make money in the stock market, guide a spacecraft, make weather forecasts, predict population growth, design buildings, quantify happiness, and analyze the spread of AIDS.

Mathematics has caused a revolution. It has shaped our thoughts. It has shaped the *way* we think. Mathematics has changed the way we look at the world.

Mathematician Micro-Biographies

The birthdays of some famous and important mathematicians are highlighted throughout this book with "Born on this day" designations. You can turn to the back of the book for microbiographies, providing just a taste of the advanced fields explored by these insightful individuals, along with the countries with which they are often associated, and a few curious facts that interest me personally about them.

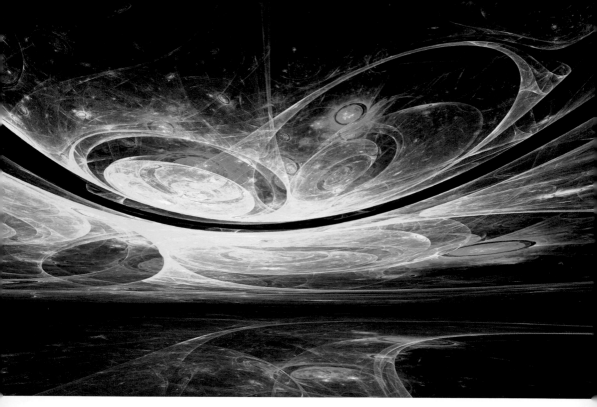

JANUARY 1

"What if I told you that you don't have to sail across an ocean
or fly into space to discover the wonders of the world? They are right here,
intertwined with our present reality. In a sense, within us, Mathematics directs
the flow of the universe, lurks beyond its shapes and curves, holds the reins
of everything from tiny atoms to the biggest stars."

—EDWARD FRENKEL, *LOVE AND MATH*, 2013

JANUARY 2

"Blindness to the aesthetic element in mathematics is widespread and can account for a feeling that mathematics is dry as dust, as exciting as a telephone book . . . Contrariwise, appreciation of this element makes the subject live in a wonderful manner and burn as no other creation of the human mind seems to do."

—PHILIP J. DAVIS AND REUBEN HERSH, *THE MATHEMATICAL EXPERIENCE*, 1981

JANUARY 3

"Number is the ruler of forms and ideas, and the cause of gods and demons."

—PYTHAGORAS, C. 500 BCE

JANUARY 4

"Mathematics, rightly viewed, possesses not only truth, but supreme beauty—
a beauty cold and austere, like that of sculpture."

—BERTRAND RUSSELL, *MYSTICISM AND LOGIC*, 1918

JANUARY 5

BORN ON THIS DAY: Camille Jordan, 1838

"We now know that there exist true propositions which we can never formally prove. What about propositions whose proofs require arguments beyond our capabilities? What about propositions whose proofs require millions of pages? Or a million, million pages? Are there proofs that are possible, but beyond us?"

—CALVIN CLAWSON, *MATHEMATICAL MYSTERIES*, 1999

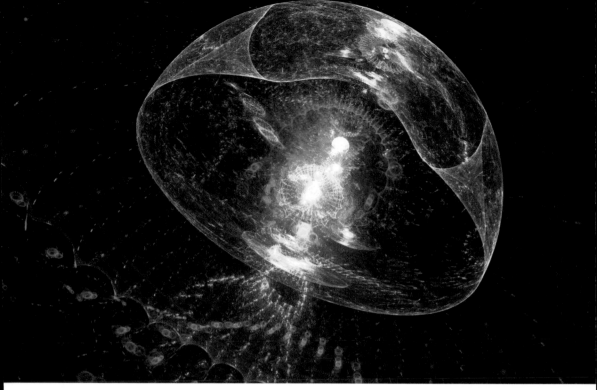

JANUARY 6

"Before creation, God did just pure mathematics.
Then He thought it would be a pleasant change to do some applied."

—JOHN EDENSOR LITTLEWOOD, *A MATHEMATICIAN'S MISCELLANY*, 1953

JANUARY 7

BORN ON THIS DAY: Émile Borel, 1871

"The trouble with integers is that we have examined only the small ones. Maybe all the exciting stuff happens at really big numbers, ones we can't get our hands on or even begin to think about in any very definite way. So maybe all the action is really inaccessible and we're just fiddling around. Our brains have evolved to get us out of the rain, find where the berries are, and keep us from getting killed. Our brains did not evolve to help us grasp really large numbers or to look at things in a hundred thousand dimensions."

—RONALD GRAHAM, QUOTED IN PAUL HOFFMAN'S
"THE MAN WHO LOVES ONLY NUMBERS," *THE ATLANTIC*, 1987

JANUARY 8

"Mathematics is as much part of our cultural heritage as art, literature, and music. As humans, we have a hunger to discover something new, reach new meaning, understand better the universe and our place in it."

—EDWARD FRENKEL, *LOVE AND MATH*, 2013

JANUARY 9

"Anyone who cannot cope with mathematics is not fully human.
At best he is a tolerable subhuman who has learned to wear shoes,
bathe, and not make messes in the house."

—ROBERT A. HEINLEIN, *TIME ENOUGH FOR LOVE*, 1973

JANUARY 10

"When mathematicians think about algorithms, it is usually from the God's-eye perspective. They are interested in proving, for instance, that *there is* some algorithm with some interesting property, or *that there is* no such algorithm, and in order to prove such things you needn't actually locate the algorithm you are talking about . . ."

—DANIEL DENNETT, *DARWIN'S DANGEROUS IDEA*, 1996

JANUARY 11

"If a lunatic scribbles a jumble of mathematical symbols it does not follow that the writing means anything merely because to the inexpert eye it is indistinguishable from higher mathematics."

—ERIC TEMPLE BELL, QUOTED IN J. R. NEWMAN'S *THE WORLD OF MATHEMATICS*, 1956

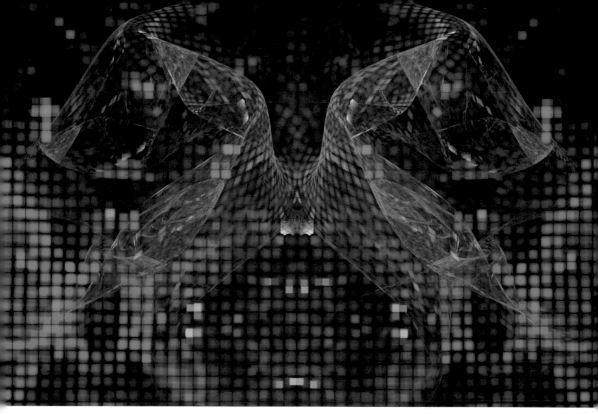

JANUARY 12

"God exists since mathematics is consistent, and the
devil exists since we cannot prove the consistency."

—MORRIS KLINE, *MATHEMATICAL THOUGHT FROM ANCIENT TO MODERN TIMES*, 1972

JANUARY 13

"Mathematics is the only infinite human activity. It is conceivable
that humanity could eventually learn everything in physics or biology. But
humanity certainly won't ever be able to find out everything in mathematics,
because the subject is infinite. Numbers themselves are infinite."

—PAUL ERDŐS, QUOTED IN PAUL HOFFMAN'S *THE MAN WHO LOVED ONLY NUMBERS*, 1998

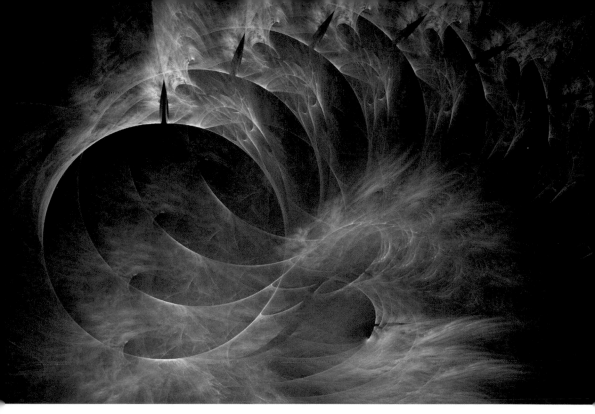

JANUARY 14

BORN ON THIS DAY: Alfred Tarski, 1902

"It is impossible to be a mathematician without being a poet in soul."

—SOFIA KOVALEVSKAYA, *RECOLLECTIONS OF CHILDHOOD*, 1895

JANUARY 15

BORN ON THIS DAY: Sofia Kovalevskaya, 1850

"Mathematical inquiry . . . lifts the human mind into closer proximity
with the divine than is attainable through any other medium."

—HERMANN WEYL, *THE OPEN WORLD*, 1932

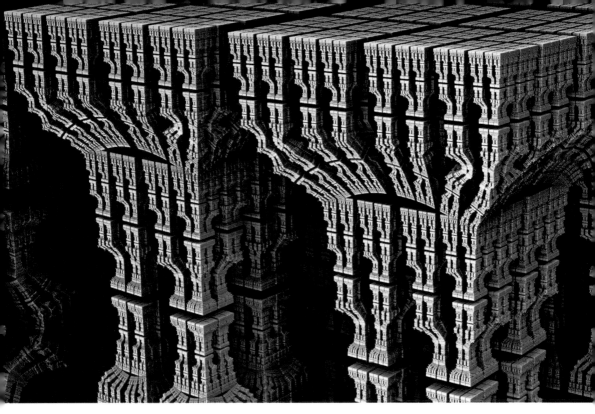

JANUARY 16

"In most sciences, one generation tears down what another has built and what one has established another undoes. In mathematics alone, each generation adds a new story to the old structure."

—HERMANN HANKEL, *DIE ENTWICKLUNG DER MATHEMATIK IN DEN LETZTEN JAHRHUNDERTEN (THE DEVELOPMENT OF MATHEMATICS IN THE LAST FEW CENTURIES)*, 1869

JANUARY 17

"Even stranger things have happened; and perhaps the strangest of all is the marvel that mathematics should be possible to a race akin to the apes."

—ERIC TEMPLE BELL, *THE DEVELOPMENT OF MATHEMATICS*, 1945

JANUARY 18

"Philosophers and great religious thinkers of the last century saw
evidence of God in the symmetries and harmonies around them—in the
beautiful equations of classical physics that describe such phenomena as electricity
and magnetism. I don't see the simple patterns underlying nature's complexity as
evidence of God. I believe that is God. To behold [mathematical curves],
spinning to their own music, is a wondrous, spiritual event."

—PAUL RAPP, QUOTED IN KATHLEEN MCAULIFFE'S "GET SMART: CONTROLLING CHAOS," *OMNI*, 1990

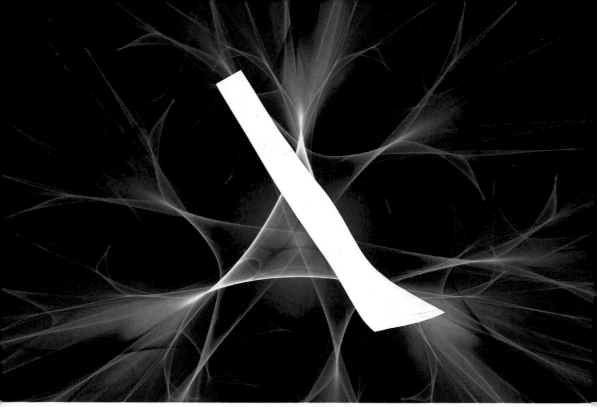

JANUARY 19

BORN ON THIS DAY: Alfred Clebsch, 1833

"Archimedes will be remembered when Aeschylus is forgotten, because languages die and mathematical ideas do not. 'Immortality' may be a silly word, but probably a mathematician has the best chance of whatever it may mean."

—G. H. HARDY, *A MATHEMATICIAN'S APOLOGY*, 1941

JANUARY 20

"Great equations change the way we perceive the world. They reorchestrate the world—transforming and reintegrating our perception by redefining what belongs together with what. Light and waves. Energy and mass. Probability and position. And they do so in a way that often seems unexpected and even strange."

—ROBERT P. CREASE, "THE GREATEST EQUATIONS EVER," *PHYSICSWEB*, 2004

JANUARY 21

"[Much of frontier mathematics] confounds even mathematicians and physicists, as they use math to calculate the inconceivable, undetectable, nonexistent and impossible. So what does it mean when mainstream explanations of our physical reality are based on stuff that even scientists cannot comprehend? When nonscientists read about the strings and branes of the latest physics theories, or the Riemann surfaces and Galois fields of higher mathematics, how close are we to a real understanding?"

—SUSAN KRUGLINSKI, "WHEN EVEN MATHEMATICIANS DON'T UNDERSTAND THE MATH," *THE NEW YORK TIMES*, 2004

JANUARY 22

"The miracle of appropriateness of the language of mathematics for the formulation of the laws of physics is a wonderful gift which we neither understand nor deserve. We should be grateful for it, and hope that it will remain valid for future research, and that it will extend, for better or for worse, to our pleasure even though perhaps also to our bafflement, to wide branches of learning."

—EUGENE WIGNER, "THE UNREASONABLE EFFECTIVENESS OF MATHEMATICS IN THE NATURAL SCIENCES," *COMMUNICATIONS ON PURE AND APPLIED MATHEMATICS*, 1960

JANUARY 23

BORN ON THIS DAY: David Hilbert, 1862

"Distance in four dimensions means nothing to the layman.
Even four-dimensional space is wholly beyond ordinary imagination.
But the mathematician is not called upon to struggle with the bounds
of imagination, but only with the limitations of his logical faculties."

—EDWARD KASNER AND JAMES NEWMAN, *MATHEMATICS AND THE IMAGINATION*, 1940

JANUARY 24

"The Gedemondan chuckles. 'We read probabilities. You see, we see—
perceive is a better word—the math of the Well of Souls. We feel the energy flow,
the ties and bands, in each and every particle of matter and energy. All reality is
mathematics, all existence—past, present, and future—is equations.'"

–JACK CHALKER, *QUEST FOR THE WELL OF SOULS*, 1978

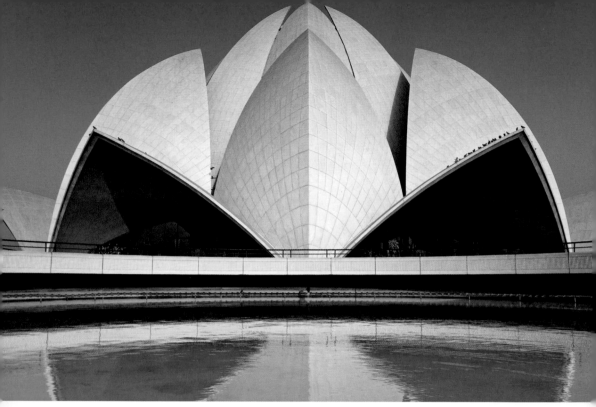

JANUARY 25

BORN ON THIS DAY: Joseph-Louis Lagrange, 1736

"Each generation has its few great mathematicians, and mathematics would not even notice the absence of the others. They are useful as teachers, and their research harms no one, but it is of no importance at all. A mathematician is great or he is nothing."

—ALFRED ADLER, "MATHEMATICS AND CREATIVITY," *THE NEW YORKER*, 1972

JANUARY 26

"I wonder whether fractal images are not touching the very structure
of our brains. Is there a clue in the infinitely regressing character of such
images that illuminates our perception of art? Could it be that a fractal image
is of such extraordinary richness, that it is bound to resonate with our
neuronal circuits and stimulate the pleasure I infer we all feel."

—PETER W. ATKINS, "ART AS SCIENCE," *THE DAILY TELEGRAPH*, 1990

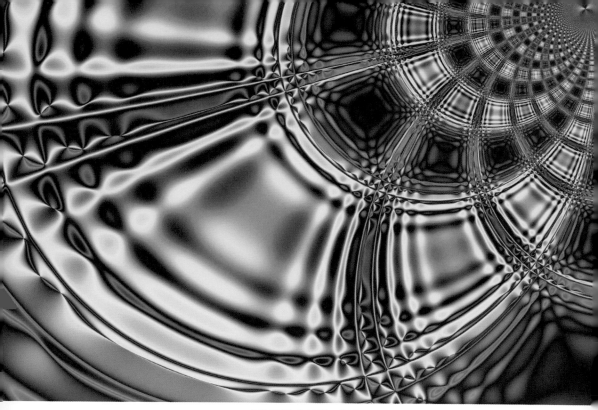

JANUARY 27

"It is a mathematical fact that the casting of this pebble from
my hand alters the center of gravity of the universe."

—THOMAS CARLYLE, *SARTOR RESARTUS*, 1831

JANUARY 28

"Music is the pleasure the human mind experiences
from counting without being aware that it is counting."

—GOTTFRIED LEIBNIZ, LETTER TO CHRISTIAN GOLDBACH, 1712

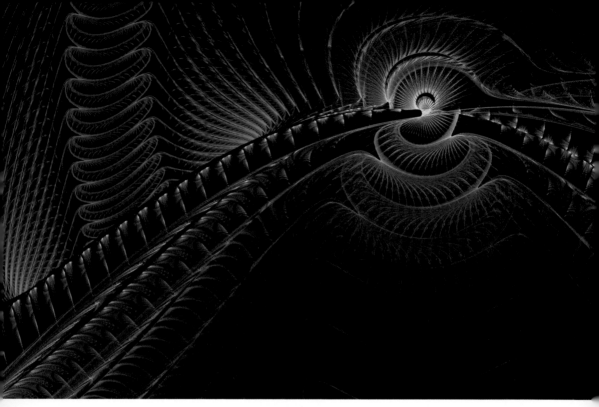

JANUARY 29

BORN ON THIS DAY: Ernst Kummer, 1810

"He remembered exploring those other-worldly curves from one degree
to the next, lemniscate to folium, progressing eventually to an ungraphable
class of curve, no precise slope at any point, a tangent-defying mind marvel."

—DON DELILLO, *RATNER'S STAR*, 1976

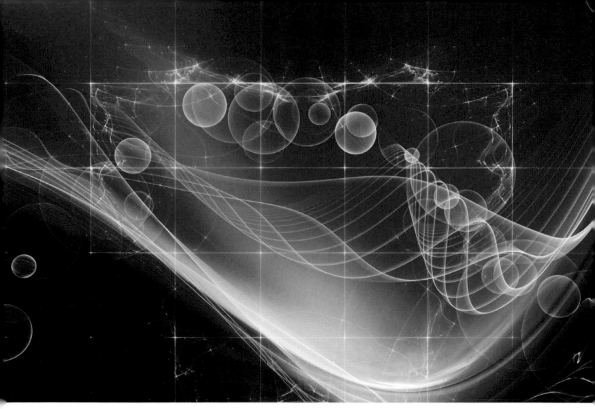

JANUARY 30

"Graphic images are the key. It's masochism for a mathematician
to do without pictures . . . [Otherwise] how can they see the relationship
between that motion and this. How can they develop intuition?"

—JAMES GLEICK, *CHAOS*, 1987

JANUARY 31

"If we wish to make a new world we have the material ready.
The first one, too, was made out of chaos."

—ROBERT QUILLEN, *OMNI*, 1987

FEBRUARY 1

"Anyone who considers arithmetical methods of
producing random digits is, of course, in a state of sin."

—JOHN VON NEUMANN, "VARIOUS TECHNIQUES USED IN CONNECTION WITH RANDOM DIGITS,"
JOURNAL OF RESEARCH OF THE NATIONAL BUREAU OF STANDARDS, 1951

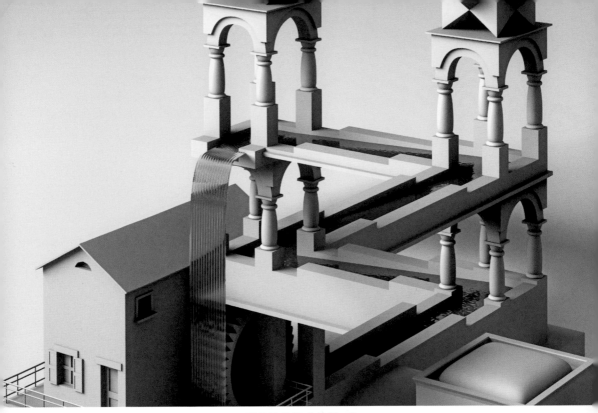

FEBRUARY 2

"Mathematics is the only science where one never knows what one is talking about nor whether what is said is true."

—BERTRAND RUSSELL, "RECENT WORK ON THE PRINCIPLES OF MATHEMATICS," *INTERNATIONAL MONTHLY*, 1901

FEBRUARY 3

"Nature is relationships in space. Geometry defines
relationships in space. Art creates relationships in space."

—M. BOLES AND R. NEWMAN, *UNIVERSAL PATTERNS*, 1990

FEBRUARY 4

"This, therefore, is mathematics: she reminds you of the invisible
forms of the soul; she gives life to her own discoveries; she awakens
the mind and purifies the intellect; she brings to light our intrinsic ideas;
she abolishes oblivion and ignorance which are ours by birth."

—PROCLUS, QUOTED IN MORRIS KLINE'S *MATHEMATICAL THOUGHT FROM ANCIENT TO MODERN TIMES*, 1990

FEBRUARY 5

"Amusement is one of humankind's strongest motivating forces. Although mathematicians sometimes belittle a colleague's work by calling it 'recreational mathematics,' much serious mathematics has come out of recreational problems, which test mathematical logic and reveal mathematical truths."

—IVARS PETERSON, *ISLANDS OF TRUTH*, 1990

FEBRUARY 6

"One sign of an interesting program is that you cannot readily predict its output."

—BRIAN HAYES, "ON THE BATHTUB ALGORITHM FOR DOT-MATRIX HOLOGRAMS," *COMPUTER LANGUAGE*, 1986

FEBRUARY 7

BORN ON THIS DAY: G. H. Hardy, 1877

"While the equations represent the discernment of eternal and universal truths, however, the manner in which they are written is strictly, provincially human. That is what makes them so much like poems, wonderfully artful attempts to make infinite realities comprehensible to finite beings."

—MICHAEL GUILLEN, *FIVE EQUATIONS THAT CHANGED THE WORLD*, 1996

FEBRUARY 8

BORN ON THIS DAY: Daniel Bernoulli, 1700

"The generation of random numbers is too important to be left to chance."

—THE TITLE OF ROBERT COVEYOU'S PAPER, WHICH APPEARED IN *STUDIES IN APPLIED MATHEMATICS*, 1969

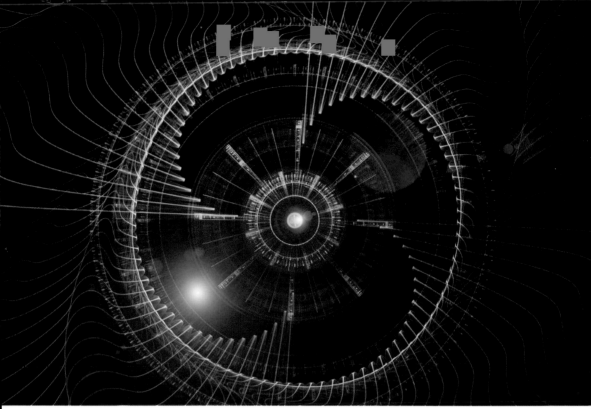

FEBRUARY 9

BORN ON THIS DAY: Harold Scott MacDonald "Donald" Coxeter, 1907

"Mathematics and music, the most sharply contrasted fields of scientific activity which can be found, and yet related, supporting each other, as if to show forth the secret connection which ties together all the activities of our mind, and which leads us to surmise that the manifestations of the artist's genius are but the unconscious expressions of a mysteriously acting rationality."

—HERMANN VON HELMHOLTZ, *VORTRÄGE UND REDEN* (*PRESENTATIONS AND SPEECHES*), 1884

FEBRUARY 10

"Applications, computers, and mathematics form a tightly coupled system yielding results never before possible and ideas never before imagined."

—LYNN ARTHUR STEEN, "THE SCIENCE OF PATTERNS," *SCIENCE*, 1988

FEBRUARY 11

"Science is not about control. It is about cultivating a perpetual condition of wonder in the face of something that forever grows one step richer and subtler than our latest theory about it. It is about reverence, not mastery."

—RICHARD POWERS, *THE GOLD BUG VARIATIONS*, 1991

FEBRUARY 12

"Look round the world: contemplate the whole and every part of it;
you will find it to be nothing but one great machine, subdivided into an
infinite number of lesser machines . . . We are led to infer that the
Author of Nature is somewhat similar to the mind of man."

–DAVID HUME, *DIALOGUES CONCERNING NATURAL RELIGION*, 1779

FEBRUARY 13

BORN ON THIS DAY: Peter Gustav Lejeune Dirichlet, 1805

"The Stone Age artist . . . relied exclusively on his mathematical instinct. It's his instinct that has been abstracted and fettered into geometrical forms. In the course of time, it helped to develop the concept of numbers . . . so that finally the manifold manifestations of being in space and time could be put in order by abstract numbers."

—ANNEMARIE SCHIMMEL, *THE MYSTERY OF NUMBERS*, 1993

FEBRUARY 14

"At all the moments of death, one lives over again his past life
with a rapidity inconceivable to others. This remembered life must also
have a last moment, and this last moment its own last moment, and so on,
and hence, dying is itself eternity, and hence, in accordance with the theory
of limits, one may approach death but can never reach it."

—ARTHUR SCHNITZLER, *FLIGHT INTO DARKNESS*, 1931

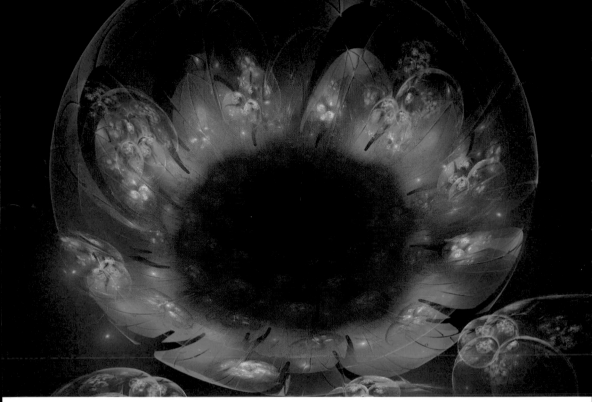

FEBRUARY 15

BORN ON THIS DAY: Alfred North Whitehead, 1861

"Even the mathematician would like to nibble the forbidden fruit, to glimpse what it would be like if he could slip for a moment into a fourth dimension."

—EDWARD KASNER AND JAMES NEWMAN, *MATHEMATICS AND THE IMAGINATION*, 1940

FEBRUARY 16

"It's like asking why Beethoven's Ninth Symphony is beautiful. If you don't see why, someone can't tell you. I know numbers are beautiful. If they aren't beautiful, nothing is."

—PAUL ERDÖS, QUOTED IN PAUL HOFFMAN'S "THE MAN WHO LOVED ONLY NUMBERS," *ATLANTIC MONTHLY*, 1987

FEBRUARY 19

"The computer is dangerously close to being our modern version
of the kaleidoscope. The twists and turns of programs give unexpected
variations of form that seem to be strikingly beautiful. But is it art? What is
beauty? Things in nature such as crystals or flowers, the human body, landscapes,
and so on, can become a meaningful part of a work of art. But when nature
is simply reflected—increasingly possible as computerized techniques
advance—its value as art becomes problematical."

—ROBERT MUELLER, *ART IN AMERICA*, 1972

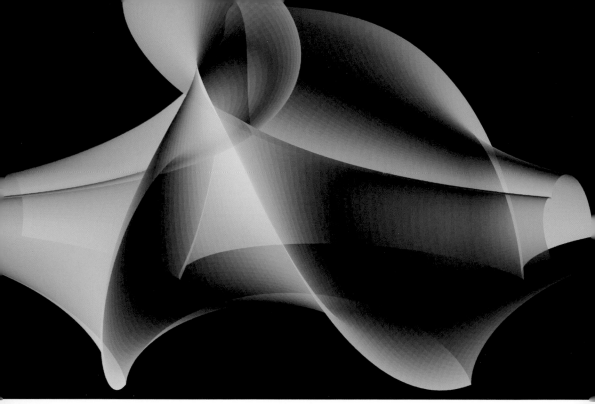

FEBRUARY 20

BORN ON THIS DAY: John Willard Milnor, 1931

"Fraa Jad took the garment from me and discovered how the fly worked. 'Topology is destiny,' he said, and put the drawers on. One leg at a time."

—NEAL STEPHENSON, *ANATHEM*, 2009

FEBRUARY 23

"A person who can within a year solve $x^2 - 92y^2 = 1$ is a mathematician."

—BRAHMAGUPTA, *THE OPENING OF THE UNIVERSE*, 628 CE

FEBRUARY 24

"The ratio of the height of the Sears Building in Chicago to the height of the Woolworth Building in New York is the same to four significant digits (1.816 vs. 1816) as the ratio of the mass of a proton to the mass of an electron."

—JOHN PAULOS, *INNUMERACY*, 1988

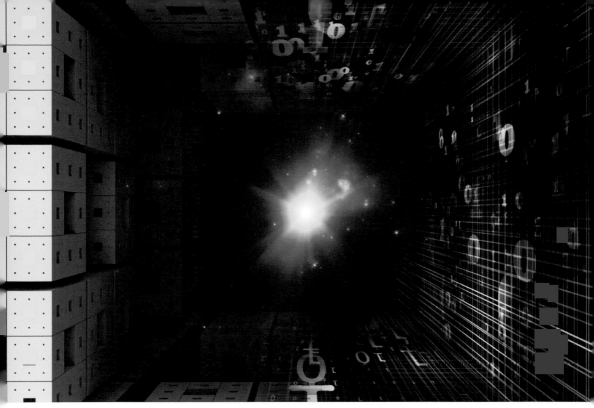

FEBRUARY 25

"Mathematicians might actually be looking at life with a most trenchant
sense—one that perceives things the other five senses cannot."

—MICHAEL GUILLEN, *BRIDGES TO INFINITY*, 1983

FEBRUARY 26

"I believe that the infinity of possibilities predicted to arise in quantum physics is the same infinity as the number of universe-possibilities predicted to arise in relativistic physics when, at the beginning of time, the universe, our home, and all of its sisters and brothers were created. As modest and troublesome as we often are, we too are nevertheless creatures of infinity."

—FRED WOLF, *PARALLEL UNIVERSES*, 1990

FEBRUARY 27

BORN ON THIS DAY: Luitzen Egbertus Jan Brouwer, 1881

"Mathematicians study structure independent of context, and their science is a voyage of exploration through all the kinds of structure and order which the human mind is capable of discerning."

—CHARLES PINTER, *A BOOK OF ABSTRACT ALGEBRA*, 1982

FEBRUARY 28

BORN ON THIS DAY: Pierre Fatou, 1878

"The bottom line for mathematicians is that the architecture has to be right. In all the mathematics that I did, the essential point was to find the right architecture. It's like building a bridge. Once the main lines of the structure are right, then the details miraculously fit. The problem is the overall design."

—FREEMAN DYSON, INTERVIEW WITH DONALD J. ALBERS, *THE COLLEGE MATHEMATICS JOURNAL*, 1994

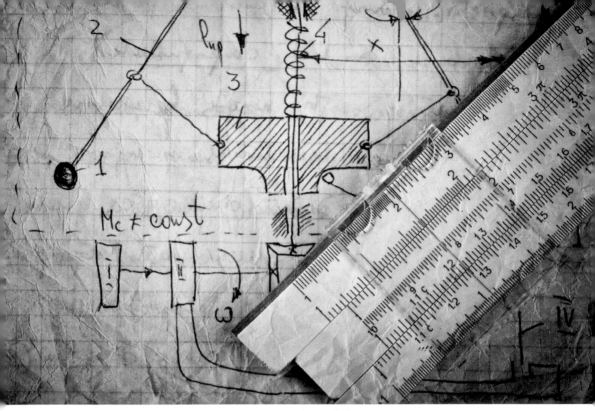

FEBRUARY 29

"Now one may ask, 'What is mathematics doing in a physics lecture?'
We have several possible excuses: first, of course, mathematics is an important
tool, but that would only excuse us for giving the formula in two minutes. On the
other hand, in theoretical physics we discover that all our laws can be written in
mathematical form; and that this has a certain simplicity and beauty about it. So,
ultimately, in order to understand nature it may be necessary to have
a deeper understanding of mathematical relationships."

—RICHARD FEYNMAN, *THE FEYNMAN LECTURES ON PHYSICS*, 1963

MARCH 1

"'Can you do addition?' the White Queen asked. 'What's one
and one and one and one and one and one and one and one and
one and one?' 'I don't know,' said Alice. 'I lost count.'"

—LEWIS CARROLL, *THROUGH THE LOOKING GLASS*, 1871

MARCH 2

BORN ON THIS DAY: Yuri Vladimirovich Matiyasevich, 1947

"Computers are composed of nothing more than logic gates stretched out to the horizon in a vast numerical irrigation system."

—STAN AUGARTEN, *STATE OF THE ART*, 1983

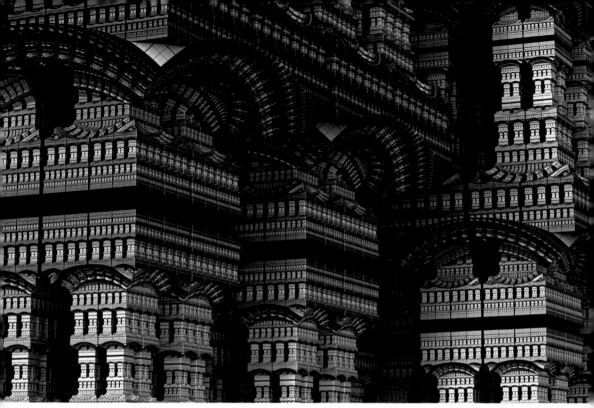

MARCH 3

BORN ON THIS DAY: Georg Cantor, 1845; Emil Artin, 1898

"Mathematics is the science of the infinite, its goal the symbolic comprehension of the infinite with human, that is finite, means. It is the great achievements of the Greeks to have made the contrast between the finite and the infinite fruitful for the cognition of reality. . . . This tension between the finite and the infinite and its conciliation now become the driving motive of Greek investigation."

—HERMANN WEYL, *THE OPEN WORLD*, 1932

MARCH 4

"In the pure mathematics we contemplate absolute truths which existed in the divine mind before the morning stars sang together, and which will continue to exist there when the last of their radiant host shall have fallen from heaven."

—EDWARD EVERETT, ADDRESS AT OPENING OF WASHINGTON UNIVERSITY, 1857

MARCH 5

"I had a feeling once about Mathematics—that I saw it all. Depth beyond depth was revealed to me—the Byss and Abyss. I saw—as one might see the transit of Venus or even the Lord Mayor's Show—a quantity passing through infinity and changing its sign from plus to minus. I saw exactly why it happened and why the tergiversation was inevitable but it was after dinner and I let it go."

—SIR WINSTON SPENCER CHURCHILL, *MY EARLY LIFE*, 1930

MARCH 6

"Perhaps an angel of the Lord surveyed an endless sea of chaos,
then troubled it gently with his finger. In this tiny and temporary
swirl of equations, our cosmos took shape."

—MARTIN GARDNER, *ORDER AND SURPRISE*, 1950

MARCH 7

"Throughout the 1960s and 1970s devoted Beckett readers greeted each successively shorter volume from the master with a mixture of awe and apprehensiveness; it was like watching a great mathematician wielding an infinitesimal calculus, his equations approaching nearer and still nearer to the null point."

—JOHN BANVILLE, *THE NEW YORK REVIEW OF BOOKS*, 1992

MARCH 8

"For a physicist, mathematics is not just a tool by means of
which phenomena can be calculated, it is the main source of concepts
and principles by means of which new theories can be created."

—FREEMAN DYSON, "MATHEMATICS IN THE PHYSICAL SCIENCES," *SCIENTIFIC AMERICAN*, 1964

MARCH 9

"It seems that if one is working from the point of view of getting beauty in one's equations, and if one has really a sound insight, one is on a sure line of progress."

—PAUL DIRAC, "THE EVOLUTION OF THE PHYSICIST'S PICTURE OF NATURE," *SCIENTIFIC AMERICAN*, 1963

MARCH 10

"The mathematician lives long and lives young; the wings of the soul do not early drop off, nor do its pores become clogged with the earthly particles blown from the dusty highways of vulgar life."

—JAMES JOSEPH SYLVESTER, PRESIDENTIAL ADDRESS TO SECTION A OF THE BRITISH ASSOCIATION, 1869

MARCH 11

"Mathematics is a field in which one's blunders tend to show very clearly and can be corrected or erased with a stroke of the pencil. It is a field which has often been compared with chess, but differs from the latter in that it is only one's best moments that count and not one's worst. A single inattention may lose a chess game, whereas a single successful approach to a problem, among many which have been relegated to the wastebasket, will make a mathematician's reputation."

—NORBERT WIENER, *EX-PRODIGY: MY CHILDHOOD AND YOUTH*, 1964

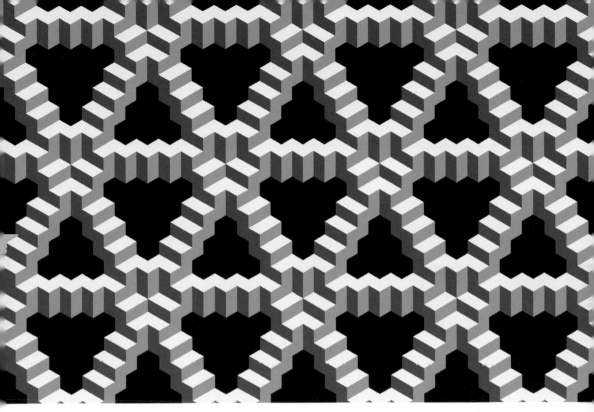

MARCH 12

"Mathematics as a science commenced when first someone, probably a Greek, proved propositions about 'any' things or about 'some' things, without specifications of definite particular things."

—ALFRED NORTH WHITEHEAD, *AN INTRODUCTION TO MATHEMATICS*, 1911

MARCH 13

"The computer generating these images was performing trillions of operations a second; in a few hours, it would manipulate more numbers than the entire human race had ever handled, since the first Cro-Magnon started counting pebbles on the floor of his cave."

—ARTHUR C. CLARKE, *THE GHOST FROM THE GRAND BANKS*, 1990

MARCH 14

"I have often been astonished to find that almost all books on philosophy and even most modern books on geometry are totally without pictures. Philosophy is supposed to be the study of thought, and I have always believed that most people thought in pictures."

—ALAN L. MACKAY, "IN THE MIND'S EYE," *COMPUTERS IN ART, DESIGN AND ANIMATION*, 1989

MARCH 15

"Science in its everyday practice is much closer to art than to philosophy. When I look at Gödel's proof of his undecidability theorem, I do not see a philosophical argument. The proof is a soaring piece of architecture, as unique and as lovely as Chartres cathedral. The proof destroyed Hilbert's dream of reducing all mathematics to a few equations, and replaced it with a greater dream of mathematics as an endlessly growing realm of ideas."

—FREEMAN DYSON, IN THE INTRODUCTION TO JOHN CORNWELL AND FREEMAN DYSON'S
NATURE'S IMAGINATION, 1995

MARCH 16

"The knowledge we have of mathematical truths is not only certain, but real knowledge; and not the bare empty vision of vain, insignificant chimeras of the brain."

—JOHN LOCKE, *AN ESSAY CONCERNING HUMAN UNDERSTANDING*, 1849

MARCH 17

"The single most compelling reason to explore the world of mathematics
is that it is beautiful, and pondering its intriguing ideas is great fun. I'm constantly
perplexed by how many people do not believe this, yet over 50,000 professional
mathematicians in America practice their trade with enthusiasm and fervor. . . .
To study the deep truths of number relationships feeds the spirit as surely as
any of the other high human activities of art, music, or literature."

—CALVIN CLAWSON, *MATHEMATICAL MYSTERIES*, 1996

MARCH 18

BORN ON THIS DAY: Christian Goldbach, 1690; Jakob Steiner, 1796

"A formal manipulator in mathematics often experiences the discomforting feeling that his pencil surpasses him in intelligence."

—HOWARD EVES, *MATHEMATICAL CIRCLES*, 1969

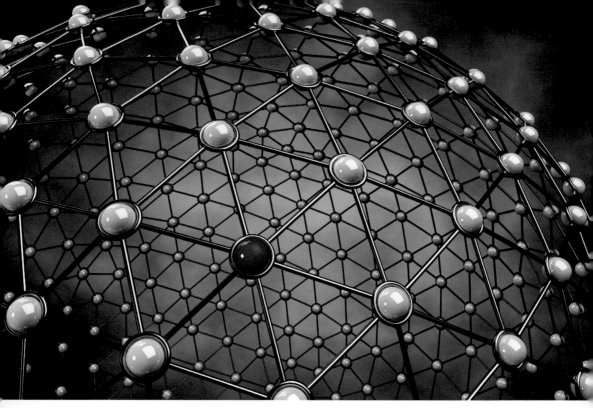

MARCH 19

"I believe there are 15,747,724,136,275,002,577,605,653,961,181,555,468,
044,717,914,527,116,709,366,231,425,076,185,631,031,296 protons
in the universe and the same number of electrons."

—SIR ARTHUR EDDINGTON, *THE PHILOSOPHY OF PHYSICAL SCIENCE*, 1939

MARCH 20

"Large numbers have a distinct appeal, a majesty if you will. In a sense, they lie at the limits of the human imagination, which is why they have long proved elusive, difficult to define, and harder still to manipulate. Modern computers now possess enough memory and speed to handle quite impressive figures. For instance, it is possible to multiply together million-digit numbers in a mere fraction of a second. As a result we can now characterize numbers about which earlier mathematicians could only dream."

—RICHARD E. CRANDALL, "THE CHALLENGE OF LARGE NUMBERS," *SCIENTIFIC AMERICAN*, 1997

MARCH 21

BORN ON THIS DAY: Joseph Fourier, 1768; George David Birkhoff, 1884

"The scientist's religious feeling takes the form of a rapturous amazement at the harmony of natural law, which reveals an intelligence of such superiority that, compared with it, all the systematic thinking and acting of human beings is an utterly insignificant reflection. This feeling is the guiding principle of his life and work . . . It is beyond question closely akin to that which has possessed the religious geniuses of all ages."

—ALBERT EINSTEIN, *MEIN WELTBILD (MY WORLDVIEW)*, 1934

MARCH 26

BORN ON THIS DAY: Paul Erdős, 1913

"There are tetradic, pandigit, and prime-factorial plus one primes.
And there are Cullen, multifactorial, beastly palindrome, and antipalindrome
primes. Add to these the strobogrammatic, subscript, internal repdigit, and the elliptic
primes. In fact, a whole new branch of mathematics seems to be evolving that deals
specifically with the attributes of the various kinds of prime numbers.
Yet understanding primes is only part of our quest to fully understand
the number sequence and all of its delightful peculiarities."

—CALVIN CLAWSON, *MATHEMATICAL MYSTERIES*, 1999

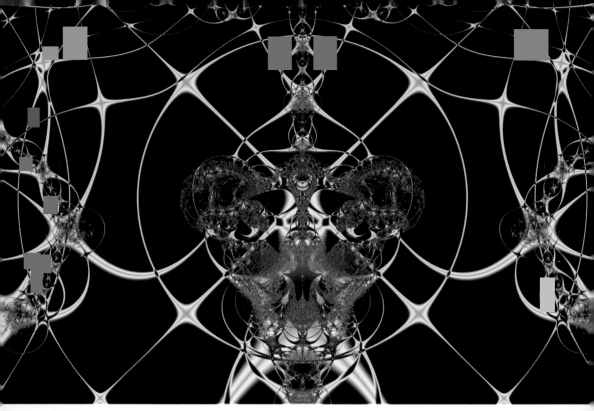

MARCH 27

"There was a blithe certainty that came from first comprehending
the full Einstein field equations, arabesques of Greek letters clinging tenuously
to the page, a gossamer web. They seemed insubstantial when you first saw them,
a string of squiggles. Yet to follow the delicate tensors as they contracted, as the
superscripts paired with subscripts, collapsing mathematically into concrete classical
entities—potential; mass; forces vectoring in a curved geometry—that was a sublime
experience. The iron fist of the real, inside the velvet glove of airy mathematics."

—GREGORY BENFORD, *TIMESCAPE*, 1980

MARCH 28

BORN ON THIS DAY: Alexander Grothendieck, 1928

"Mathematics is not a careful march down a well-cleared highway,
but a journey into a strange wilderness, where the explorers often get lost.
Rigor should be a signal to the historian that the maps have been made,
and the real explorers have gone elsewhere."

—W. S. ANGLIN, "MATHEMATICS AND HISTORY," *MATHEMATICAL INTELLIGENCER*, 1992

MARCH 29

"The higher arithmetic presents us with an inexhaustible storehouse of interesting truths—of truths, too, which are not isolated, but stand in the closest relation to one another, and between which, with each successive advance of the science, we continually discover new and wholly unexpected points of contact. A great part of the theories of arithmetic derive an additional charm from the peculiarity that we easily arrive by induction at important propositions, which have the stamp of simplicity upon them, but the demonstration of which lies so deep as not to be discovered until after many fruitless efforts; and even then it is obtained by some tedious and artificial process, while the simpler methods of proof long remain hidden from us."

—CARL FRIEDRICH GAUSS, INTRODUCTION TO GOTTHOLD EISENSTEIN'S
MATHEMATISCHE ABHANDLUNGEN (MATHEMATICAL TREATISES), 1849

MARCH 30

"I tell them that if they will occupy themselves with the study of mathematics they will find in it the best remedy against the lusts of the flesh."

—THOMAS MANN, *THE MAGIC MOUNTAIN*, 1924

MARCH 31

BORN ON THIS DAY: René Déscartes, 1596

"Laws of physics and mathematics are like a coordinate system that runs
in only one dimension. Perhaps there is another dimension perpendicular to it,
invisible to those laws of physics, describing the same things with different rules,
and those rules are written in our hearts, in a deep place where we
cannot go and read them except in our dreams."

—NEAL STEPHENSON, *THE DIAMOND AGE*, 1995

APRIL 1

"A good notation has a subtlety and suggestiveness which,
at times, make it almost seem like a live teacher."

—BERTRAND RUSSELL, INTRODUCTION TO LUDWIG WITTGENSTEIN'S *TRACTATUS
LOGICO-PHILOSOPHICUS (LOGICAL-PHILOSOPHICAL TREATISE)*, 1922

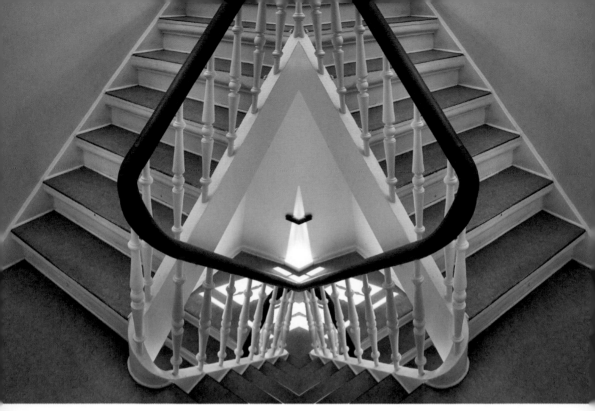

APRIL 2

BORN ON THIS DAY: Paul Cohen, 1934

"I advise my students to listen carefully the moment they decide to take no more mathematics courses. They might be able to hear the sound of closing doors."

—JAMES CABALLERO, "EVERYBODY A MATHEMATICIAN?," *CAIP QUARTERLY*, 1989

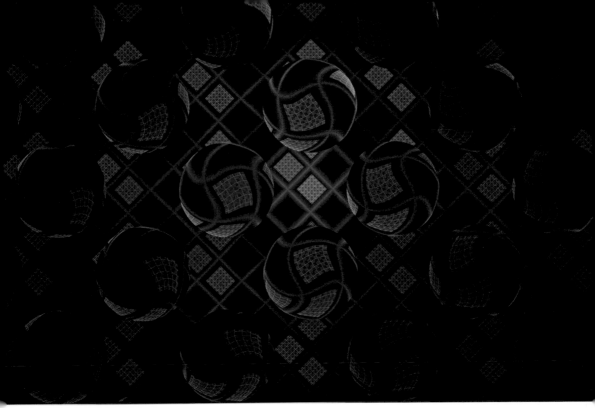

APRIL 3

"Imagine that the world is something like a great chess game being
played by the gods, and we are observers of the game. . . . If we watch long enough,
we may eventually catch on to a few of the rules. . . . However, we might not be able
to understand why a particular move is made in the game, merely because it is too
complicated and our minds are limited. . . . We must limit ourselves to the
more basic question of the rules of the game. If we know the rules,
we consider that we 'understand' the world."

—RICHARD FEYNMAN, *THE FEYNMAN LECTURES ON PHYSICS*, 1963

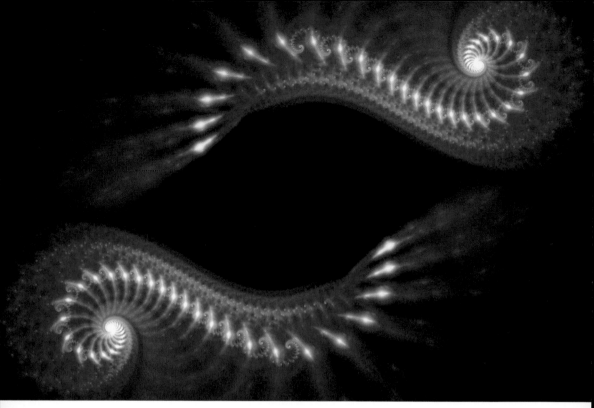

APRIL 4

"'No one really understood music unless he was a scientist,' her father had declared, and not just a scientist, either, oh, no, only the real ones, the theoreticians, whose language is mathematics. She had not understood mathematics until he had explained to her that it was the symbolic language of relationships. 'And relationships,' he had told her, 'contained the essential meaning of life.'"

—PEARL S. BUCK, *THE GODDESS ABIDES*, 1973

APRIL 5

"In the company of friends, writers can discuss their books, economists
the state of the economy, lawyers their latest cases, and businessmen their latest
acquisitions, but mathematicians cannot discuss their mathematics at all.
And the more profound their work, the less understandable it is."

—ALFRED ADLER, "REFLECTIONS: MATHEMATICS AND CREATIVITY," *THE NEW YORKER*, 1972

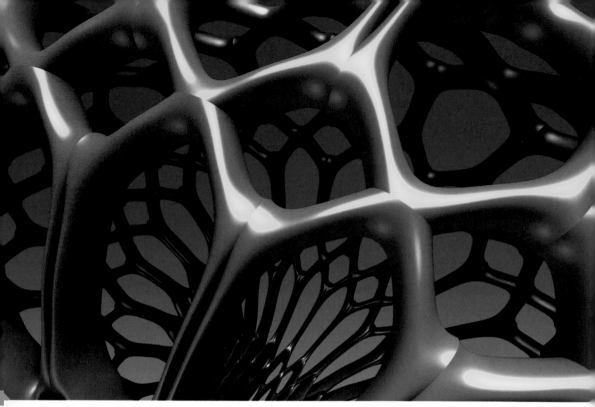

APRIL 6

"Models in the mathematical, physical and mechanical sciences are
of the greatest importance. Long ago philosophy perceived the essence of our
process of thought to lie in the fact that we attach to the various real objects around
us particular physical attributes—our concepts—and by means of these try to represent
the objects to our minds. . . . On this view our thoughts stand to things
in the same relation as models to the objects they represent."

—LUDWIG BOLTZMANN, *ENCYCLOPAEDIA BRITANNICA*, 1902

APRIL 7

"Almost your entire mathematical life has been spent on the real line
and in real space working with real numbers. Some have dipped into complex
numbers, which are just the real numbers after you throw in i. Are these the only
numbers that can be built from the rationals? The answer is no. There are entire
parallel universes of number that are totally unrelated to the real and complex
numbers. Welcome to the world of p-adic analysis—where arithmetic replaces
the tape measure and numbers take on a whole new look."

—EDWARD BURGER, "EXPLORING P-ADIC NUMBERS," DEPARTMENT OF MATHEMATICS,
KANSAS STATE UNIVERSITY WEBSITE, 2013

APRIL 8

"Mathematics is not a book confined within a cover and bound between brazen clasps, whose contents it needs only patience to ransack; it is not a mine, whose treasures may take long to reduce into possession, but which fill only a limited number of veins and lodes; it is not a soil, whose fertility can be exhausted by the yield of successive harvests; it is not a continent or an ocean, whose area can be mapped out and its contour defined: . . . its possibilities are as infinite as the worlds which are forever crowding in and multiplying upon the astronomer's gaze; it is as incapable of being restricted within assigned boundaries or being reduced to definitions of permanent validity, as the consciousness, the life, which seems to slumber in each monad, in every atom of matter, in each leaf and bud and cell, and is forever ready to burst forth into new forms of vegetable and animal existence."

—JAMES JOSEPH SYLVESTER, ADDRESS ON COMMEMORATION DAY AT JOHNS HOPKINS UNIVERSITY, 1877

APRIL 9

BORN ON THIS DAY: George Peacock, 1791; Élie Joseph Cartan, 1869

"The great equations of modern physics are a permanent part of scientific knowledge, which may outlast even the beautiful cathedrals of earlier ages."

—STEVEN WEINBERG, IN GRAHAM FARMELO'S *IT MUST BE BEAUTIFUL*, 2003

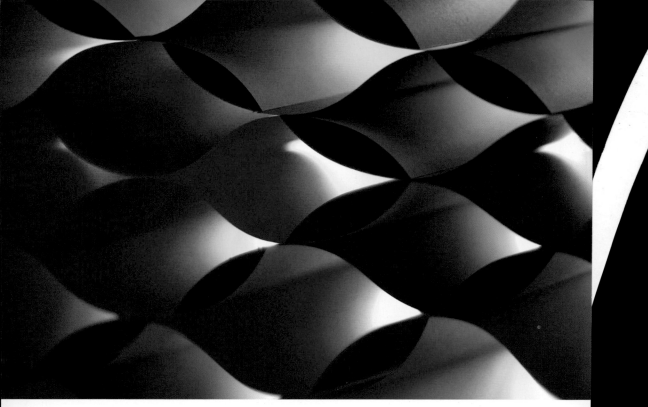

APRIL 10

"Numbers written on restaurant bills within the confines of restaurants do not follow the same mathematical laws as numbers written on any other pieces of paper in any other parts of the Universe. This single statement took the scientific world by storm. It completely revolutionized it. So many mathematical conferences got held in such good restaurants that many of the finest minds of a generation died of obesity and heart failure and the science of math was put back by years."

—DOUGLAS ADAMS, *LIFE, THE UNIVERSE AND EVERYTHING*, 1982

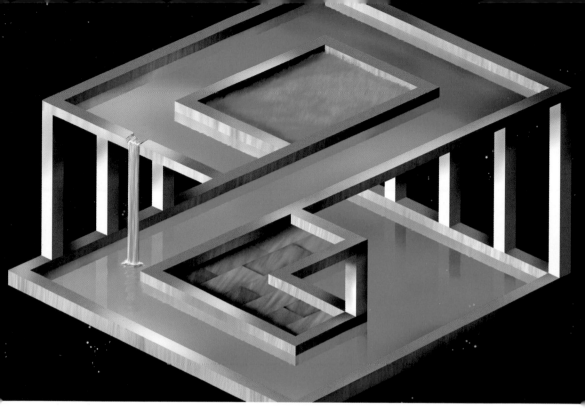

APRIL 11

BORN ON THIS DAY: Andrew John Wiles, 1953

"At this point an enigma presents itself which in all ages has agitated inquiring minds. How can it be that mathematics, being after all a product of human thought which is independent of experience, is so admirably appropriate to the objects of reality? Is human reason, then, without experience, merely by taking thought, able to fathom the properties of real things? In my opinion the answer to this question is briefly this: As far as the laws of mathematics refer to reality, they are not certain; and as far as they are certain, they do not refer to reality."

—ALBERT EINSTEIN, ADDRESS TO THE PRUSSIAN ACADEMY OF SCIENCES IN BERLIN, 1921

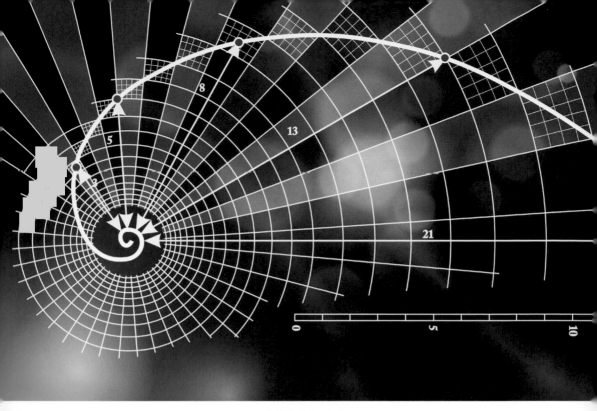

APRIL 12

"Our mathematical models of physical reality are far from complete, but they provide us with schemes that model reality with great precision—a precision enormously exceeding that of any description that is free of mathematics."

—ROGER PENROSE, "WHAT IS REALITY?," *NEW SCIENTIST*, 2006

APRIL 15

BORN ON THIS DAY: Leonhard Euler, 1707; Hermann Grassmann, 1809

"It can be of no practical use to know that pi is irrational, but if we can know, it surely would be intolerable not to know. Pure mathematicians do Mathematics because it gives them an aesthetic satisfaction which they can share with other mathematicians. They do it because for them it is fun, in the same way perhaps that people climb mountains for fun. It may be an extremely arduous and even fatal pursuit, but it is fun nevertheless. Mathematicians enjoy themselves because they do sometimes get to the top of their mountains, and anyhow trying to get up does seem to be worthwhile."

—EDWARD C. TITCHMARSH, *MATHEMATICS FOR THE GENERAL READER*, 1948

APRIL 16

BORN ON THIS DAY: Gotthold Eisenstein, 1823

"As measured by the millions of those who speak it fluently . . . , mathematics is arguably the most successful global language ever spoken. . . . Equations are like poetry: They speak truths with a unique precision, convey volumes of information in rather brief terms . . . And just as conventional poetry helps us to see deep within ourselves, mathematical poetry helps us to see far beyond ourselves."

—MICHAEL GUILLEN, *FIVE EQUATIONS THAT CHANGED THE WORLD*, 1995

APRIL 17

"Classical mathematics had its roots in the regular geometrical structures
of Euclid and Newton. Modern mathematics began with Cantor's set theory and
Peano's space-filling curve. . . . These new structures were regarded as 'pathological,'
as a 'gallery of monsters,' kin to the cubist painting and atonal music that were
upsetting established standards. . . . The mathematicians who created the monsters
regard them as important in showing that the world of pure mathematics contains a
richness of possibilities going far beyond simple structures that they saw in Nature. . . .
But Nature has played a joke on the mathematicians. . . . The same pathological
structures that the mathematicians invented to break loose from 19th-century
naturalism turned out to be inherent in familiar objects all around us."

—FREEMAN DYSON, "CHARACTERIZING IRREGULARITY," *SCIENCE*, 1978

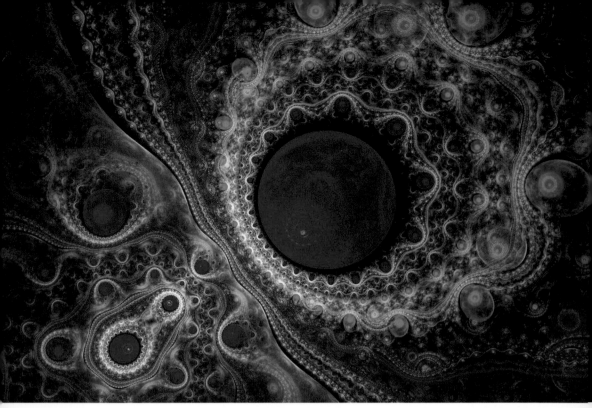

APRIL 18

BORN ON THIS DAY: Brook Taylor, 1685

"There is no excellent beauty that hath not some strangeness in the proportion."

—FRANCIS BACON, "OF BEAUTY," 1625

APRIL 21

BORN ON THIS DATE: Michael Freedman, 1951

"What could be more beautiful than a deep, satisfying relation between whole numbers? How high they rank, in the realms of pure thought and aesthetics, above their lesser brethren: the real and complex numbers . . ."

—MANFRED SCHROEDER, *NUMBER THEORY IN SCIENCE AND COMMUNICATION*, 1984

APRIL 22

BORN ON THIS DAY: Sir Michael Atiyah, 1929

"In any case, do you really think kids even want something that is relevant to their daily lives? You think something practical like compound interest is going to get them excited? People enjoy fantasy, and that is just what mathematics can provide—a relief from daily life, an anodyne to the practical workaday world."

–PAUL LOCKHART, *A MATHEMATICIAN'S LAMENT*, 2009

APRIL 23

"An agnostic friend of mine once was so struck by [a recursive plot's]
infinitely many infinities that he called it 'a picture of God,'
which I don't think is blasphemous at all."

—DOUGLAS HOFSTADTER, *GÖDEL, ESCHER, BACH*, 1979

APRIL 24

"The pages and pages of complex, impenetrable calculations might have contained the secrets of the universe, copied out of God's notebook. In my imagination, I saw the creator of the universe sitting in some distant corner of the sky, weaving a pattern of delicate lace so fine that even the faintest light would shine through it. The lace stretches out infinitely in every direction, billowing gently in the cosmic breeze. You want desperately to touch it, hold it up to the light, rub it against your cheek. And all we ask is to be able to re-create the pattern, weave it again with numbers, somehow, in our own language; to make the tiniest fragment our own, to bring it back to earth."

—YÔKO OGAWA, *THE HOUSEKEEPER AND THE PROFESSOR*, 2009

APRIL 25

BORN ON THIS DAY: Felix Klein, 1849; Andrey Kolmogorov, 1903

"It was formerly supposed that Geometry was the study of the nature of the space in which we live, and accordingly it was urged, by those who held that what exists can only be known empirically, that Geometry should really be regarded as belonging to applied mathematics. But it has gradually appeared, by the increase of non-Euclidean systems, that Geometry throws no more light upon the nature of space than arithmetic throws upon the population of the United States."

—BERTRAND RUSSELL, "MATHEMATICS AND METAPHYSICIANS,"
MYSTICISM AND LOGIC AND OTHER ESSAYS, 1918

APRIL 26

"A science of all these possible kinds of space would undoubtedly be
the highest enterprise which a finite understanding could undertake in the field of
geometry. . . . If it is possible that there could be regions with other dimensions, it is
very likely that a God had somewhere brought them into being. Such higher spaces
would not belong to our world, but form separate worlds."

—IMMANUEL KANT, *THOUGHTS ON THE TRUE ESTIMATION OF LIVING FORCES***, 1747**

APRIL 27

"The progress of mathematics and physics impels us to fly away on the wings of the poetic imagination out beyond the frontiers of Euclidean space, and to attempt to conceive of space in which more than three coordinates can stand perpendicularly to one another. But all such endeavors to fly out beyond our frontiers always end with our falling back with singed wings on the ground of our Euclidean three-dimensional space. If we attempt to contemplate the fourth dimension we encounter an insurmountable obstacle. . . . We can certainly *calculate* with [higher-dimensional spaces]. But we cannot *conceive* of them. We are confined within the space in which we find ourselves when we enter into our existence, as though in a prison. Two-dimensional beings can believe in a third dimension. But they cannot *see* it."

—KARL HEIM, *CHRISTIAN FAITH AND NATURAL SCIENCE*, 1952

APRIL 28

BORN ON THIS DAY: Kurt Gödel, 1906

"Gödel proved that the world of pure mathematics is inexhaustible; no finite set of axioms and rules of inference can ever encompass the whole of mathematics; given any finite set of axioms, we can find meaningful mathematical questions which the axioms leave unanswered. I hope that an analogous situation exists in the physical world. If my view of the future is correct, it means that the world of physics and astronomy is also inexhaustible; no matter how far we go into the future, there will always be new things happening, new information coming in, new worlds to explore, a constantly expanding domain of life, consciousness, and memory."

—FREEMAN DYSON, "TIME WITHOUT END," *REVIEWS OF MODERN PHYSICS*, 1979

122

APRIL 29

BORN ON THIS DAY: Jules Henri Poincaré, 1854

"All mathematicians live in two different worlds. They live in a crystalline world of perfect platonic forms. An ice palace. But they also live in the common world where things are transient, ambiguous, subject to vicissitudes. Mathematicians go backward and forward from one world to another. They're adults in the crystalline world, infants in the real one."

—SYLVAIN CAPPELL, QUOTED IN SYLVIA NASAR'S *A BEAUTIFUL MIND*, 1998

APRIL 30

BORN ON THIS DAY: Carl Friedrich Gauss, 1777

"Mathematics is the gate and key of the sciences . . . Neglect of mathematics works injury to all knowledge, since he who is ignorant of it cannot know the other sciences or the things of this world. And what is worse, men who are thus ignorant are unable to perceive their own ignorance and so do not seek a remedy."

—ROGER BACON, *OPUS MAJUS*, 1266

MAY 1

"Cantor was careful to stress that despite the actual infinite nature of the universe, and the reasonableness of his conjecture that corporeal and aetherical monads were related to each other as powers equivalent to transfinite cardinals $\aleph_0$ and $\aleph_1$, this did not mean that God necessarily had to create worlds in this way."

—JOSEPH DAUBEN, *GEORG CANTOR*, 1979

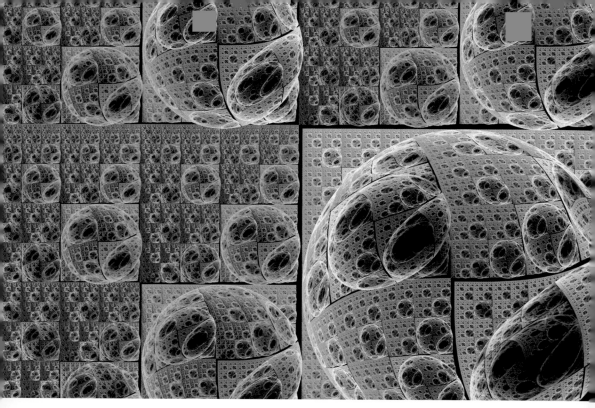

MAY 2

"I guess I think of lotteries as a tax on the mathematically challenged."

—ROGER JONES, "'I'LL THROW IN 5 BUCKS,' AND RECORD PRIZE IS CLAIMED,"
THE NEW YORK TIMES, 1998

MAY 3

"Nothing comforted Sabine like long division. That was how she had passed time waiting for Phan and then Parsifal to come back from their tests. She figured the square root of the date while other people knit and read. Sabine blamed much of the world's unhappiness on the advent of calculators."

—ANN PATCHETT, *THE MAGICIAN'S ASSISTANT*, 1998

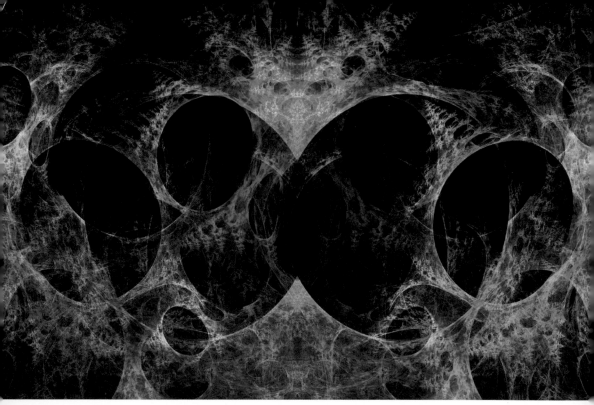

MAY 4

William Kingdon Clifford, 1845

"The important thing to remember about mathematics is not to be frightened."

—RICHARD DAWKINS, *THE BLIND WATCHMAKER*, 1986

MAY 5

"I abandoned the assigned problems in standard calculus textbooks and followed my curiosity. Wherever I happened to be—a Vegas casino, Disneyland, surfing in Hawaii, or sweating on the elliptical in Boesel's Green Microgym— I asked myself, 'Where is the calculus in this experience?'"

—JENNIFER OUELLETTE, *THE CALCULUS DIARIES*, 2010

MAY 6

BORN ON THIS DAY: André Weil, 1906

"Proof is an idol before which the mathematician tortures himself."

—SIR ARTHUR EDDINGTON, *THE NATURE OF THE PHYSICAL WORLD*, 1930

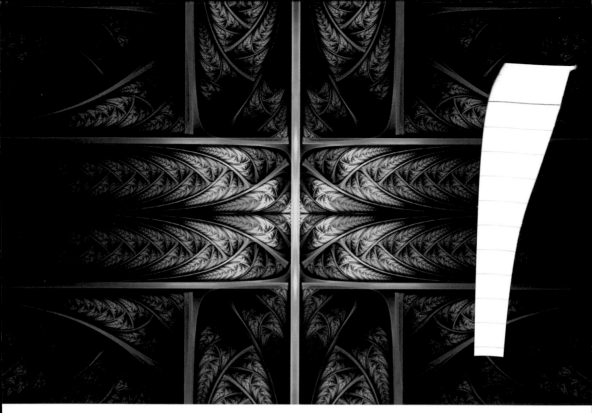

MAY 7

"Mathematics is order and beauty at its purest,
order that transcends the physical world."

—PAUL HOFFMAN, *THE MAN WHO LOVED ONLY NUMBERS*, 1998

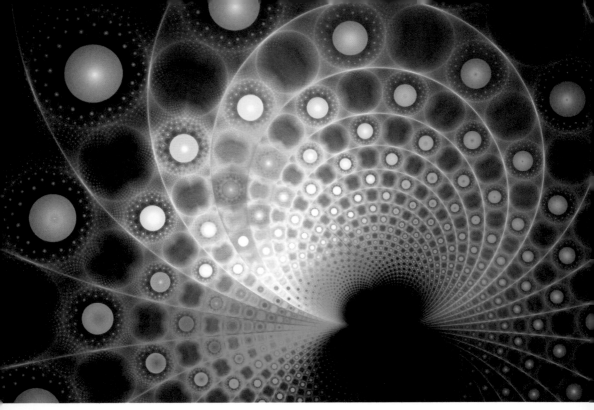

MAY 8

"Like works of literature, mathematical ideas help expand our circle
of empathy, liberating us from the tyranny of a single, parochial point of
view. Numbers, properly considered, make us better people."

—DANIEL TAMMET, *THINKING IN NUMBERS*, 2013

MAY 9

BORN ON THIS DAY: Gaspard Monge, 1746

"The advancement and perfection of mathematics are intimately connected with the prosperity of the State."

—NAPOLÉON BONAPARTE, *CORRESPONDANCE DE NAPOLÉON*, 1868

MAY 10

"We are in the position of a little child entering a huge library whose walls are covered to the ceiling with books in many different tongues. . . . The child does not understand the languages in which they are written. He notes a definite plan in the arrangement of books, a mysterious order which he does not comprehend, but only dimly suspects."

—ALBERT EINSTEIN, INTERVIEWED IN G. S. VIERECK'S *GLIMPSES OF GREAT*, 1930

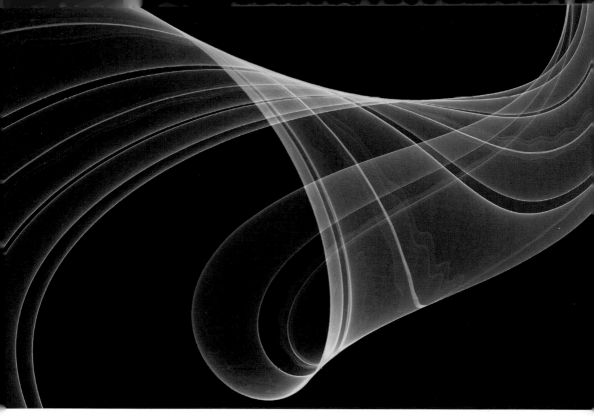

MAY 11

"We must admit with humility that, while number is purely a product of our minds, space has a reality outside our minds, so that we cannot completely prescribe its properties a priori."

—CARL FRIEDRICH GAUSS, LETTER TO FRIEDRICH BESSEL, 1830

MAY 12

"It would be very discouraging if somewhere down the line you could ask a computer if the Riemann hypothesis is correct and it said, 'Yes, it is true, but you won't be able to understand the proof.'"

—RONALD GRAHAM, "THE DEATH OF PROOF," *SCIENTIFIC AMERICAN*, 1993

MAY 13

BORN ON THIS DAY: Alexis Clairaut, 1713

"There is a noble vision of the great Castle of Mathematics, towering somewhere in the Platonic World of Ideas, which we humbly and devotedly discover (rather than invent). The greatest mathematicians manage to grasp outlines of the Grand Design, but even those to whom only a pattern on a small kitchen tile is revealed, can be blissfully happy. . . . Mathematics is a proto-text whose existence is only postulated but which nevertheless underlies all corrupted and fragmentary copies we are bound to deal with. The identity of the writer of this proto-text (or of the builder of the Castle) is anybody's guess. . . ."

—YURI I. MANIN, "MATHEMATICAL KNOWLEDGE: INTERNAL, SOCIAL, AND CULTURAL ASPECTS," *MATHEMATICS AS METAPHOR: SELECTED ESSAYS,* 2007

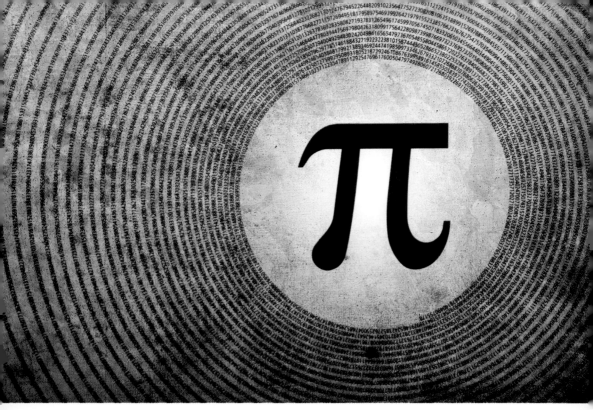

MAY 14

"The digits of pi beyond the first few decimal places are of no practical
or scientific value. Four decimal places are sufficient for the design of the finest
engines; ten decimal places are sufficient to obtain the circumference of the earth
within a fraction of an inch if the earth were a smooth sphere."

—PETR BECKMANN, *A HISTORY OF PI*, 1976

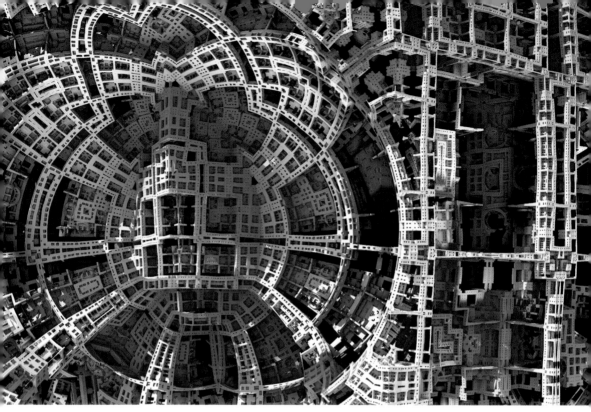

MAY 15

"When you discover mathematical structures that you believe correspond to the world around you . . . you are communicating with the universe, seeing beautiful and deep structures and patterns that no one without your training can see. The mathematics is there, it's leading you, and you are discovering it. Mathematics is a profound language, an awesomely beautiful language. For some, like Leibniz, it is the language of God. I'm not religious, but I do believe that the universe is organized mathematically."

—ANTHONY TROMBA, "UCSC PROFESSOR SEEKS TO RECONNECT MATHEMATICS TO ITS INTELLECTUAL ROOTS," UNIVERSITY OF CALIFORNIA PRESS RELEASE, 2003

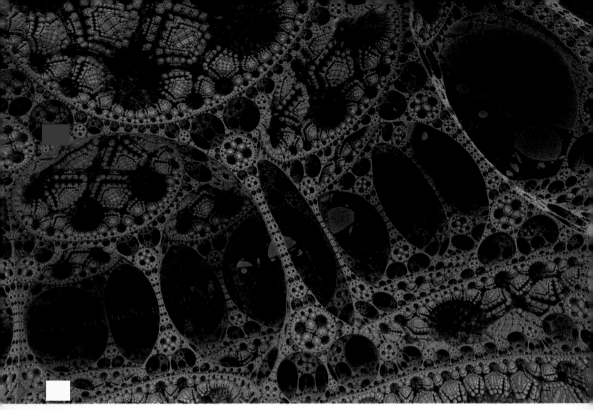

MAY 16

BORN ON THIS DAY: Maria Gaetana Agnesi, 1718; Pafnuty Chebyshev, 1821

"Mathematics as we know it and as it has come to shape modern science could never have come into being without some disregard for the dangers of the infinite."

—DAVID BRESSOUD, *A RADICAL APPROACH TO REAL ANALYSIS*, 2007

MAY 17

"The peculiar interest of [mathematical] magic squares lies in the fact that
they possess the charm of mystery. They appear to betray some hidden intelligence
which by a preconceived plan produces the impression of intentional design,
a phenomenon which finds its close analogue in nature."

—PAUL CARUS, IN W. S. ANDREWS'S *MAGIC SQUARES AND CUBES*, 1917

MAY 18

BORN ON THIS DAY: Omar Khayyám, 1048; Bertrand Russell, 1872

"As the island of knowledge grows, the surface that makes contact with mystery expands. When major theories are overturned, what we thought was certain knowledge gives way, and knowledge touches upon mystery differently. This newly uncovered mystery may be humbling and unsettling, but it is the cost of truth. Creative scientists, philosophers, and poets thrive at this shoreline."

—W. MARK RICHARDSON, "A SKEPTIC'S SENSE OF WONDER," *SCIENCE*, 1998

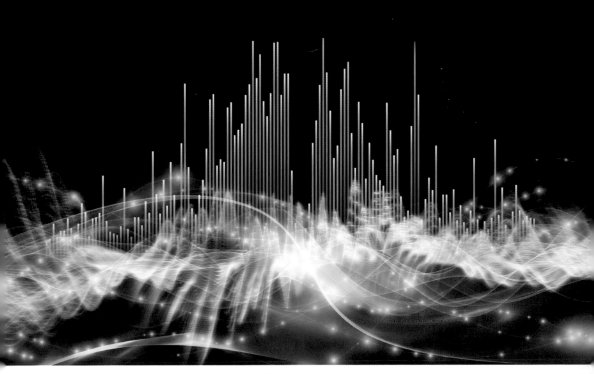

MAY 19

"The pure mathematician, like the musician, is a free
creator of his world of ordered beauty."

—BERTRAND RUSSELL, *A HISTORY OF WESTERN PHILOSOPHY*, 1945

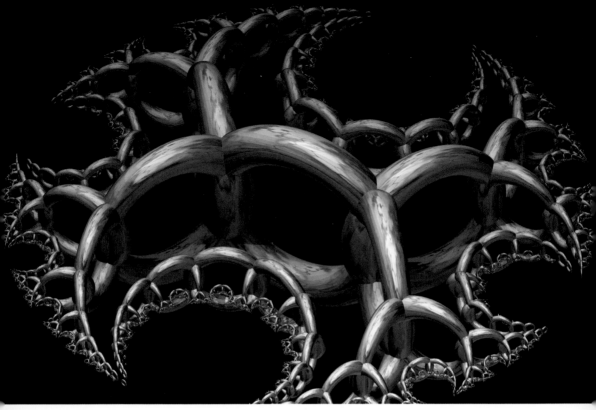

MAY 20

"The chamois making a giant leap from rock to rock and alighting,
with its full weight, on hooves supported by an ankle two centimeters in diameter:
that is challenge and that is mathematics. The mathematical phenomenon always
develops out of simple arithmetic, so useful in everyday life, out of numbers, those
weapons of the gods: the gods are there, behind the wall, at play with numbers."

—LE CORBUSIER, *THE MODULOR,* **1954**

MAY 21

"Magic squares are conspicuous instances of the intrinsic harmony of number, and so they will serve as an interpreter of the cosmic order that dominates all existence. Though they are a mere intellectual play, they not only illustrate the nature of mathematics but also the nature of existence dominated by mathematical regularity."

—PAUL CARUS, IN W. S. ANDREWS'S *MAGIC SQUARES AND CUBES*, 1917

MAY 22

"The physical method becomes a philosophy when it asserts there is no higher knowledge than the empirical knowledge of scientific phenomena. The mathematical method becomes a philosophy when it asserts that some higher knowledge is needed to explain scientific facts, and that higher knowledge is mathematics."

—FULTON J. SHEEN, *PHILOSOPHY OF SCIENCE*, 1934

MAY 23

BORN ON THIS DAY: Edward Norton Lorenz, 1917

"Our experience hitherto justifies us in believing that nature is the realization
of the simplest conceivable mathematical ideas. I am convinced that we can
discover by purely mathematical constructions the concepts and the laws connecting
[mathematics and physical reality] with each other, which furnish the key to the
understanding of natural phenomena. . . . In a certain sense, therefore, I hold
it true that pure thought can grasp reality, as the ancients dreamed."

—ALBERT EINSTEIN, "ON THE METHODS OF THEORETICAL PHYSICS," *IDEAS AND OPINIONS*, 1933

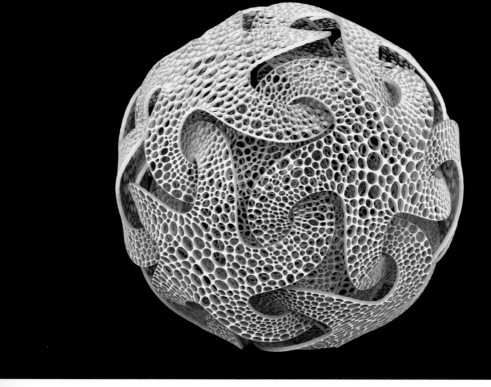

MAY 24

"As a teacher, Tengo pounded into his students' heads how voraciously mathematics demanded logic. Here things that could not be proven had no meaning, but once you had succeeded in proving something, the world's riddles settled into the palm of your hand like a tender oyster."

—HARUKI MURAKAMI, *1Q84*, 2011

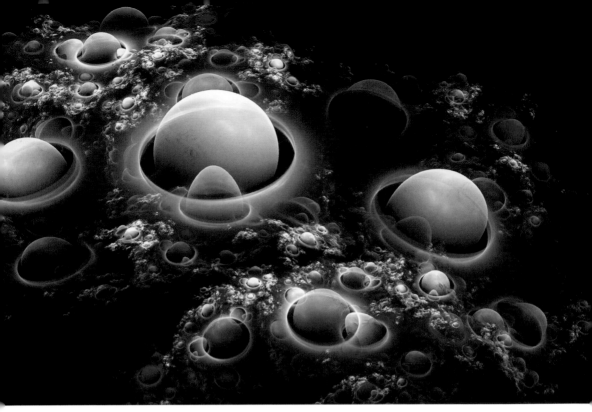

MAY 25

"The new mathematics is a sort of supplement to language, affording a means of thought about form and quantity and a means of expression, more exact, compact, and ready than ordinary language. The great body of physical science, a great deal of the essential fact of financial science, and endless social and political problems are only accessible and only thinkable to those who have had a sound training in mathematical analysis, and the time may not be very remote when it will be understood that for complete initiation as an efficient citizen of one of the new great complex world wide states that are now developing, it is as necessary to be able to compute, to think in averages and maxima and minima, as it is now to be able to read and write."

—H. G. WELLS, *MANKIND IN THE MAKING*, 1906

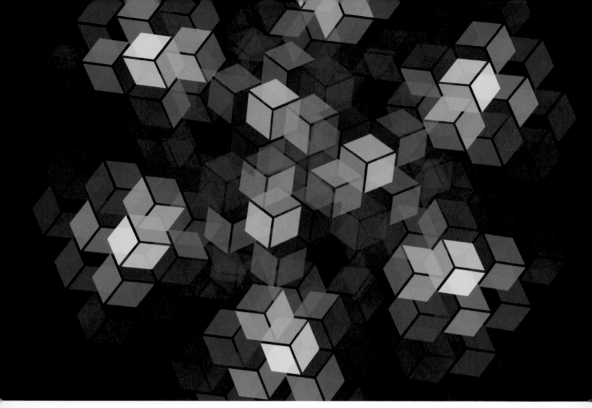

MAY 26

BORN ON THIS DAY: Abraham de Moivre, 1667

"Misunderstanding of probability may be the greatest
of all impediments to scientific literacy."

—STEPHEN JAY GOULD, *DINOSAUR IN A HAYSTACK*, 1996

MAY 27

"Heisenberg once made the following remark to Einstein:
'If nature leads us to mathematical forms of great simplicity and beauty . . .
that no one has previously encountered, we cannot help thinking that
they are "true," that they reveal a genuine feature of nature.'"

—PAUL DAVIES, *SUPERFORCE*, 1984

MAY 28

"Physics depends on a universe infinitely centered on an equals sign."

—MARK Z. DANIELEWSKI, *HOUSE OF LEAVES*, 2000

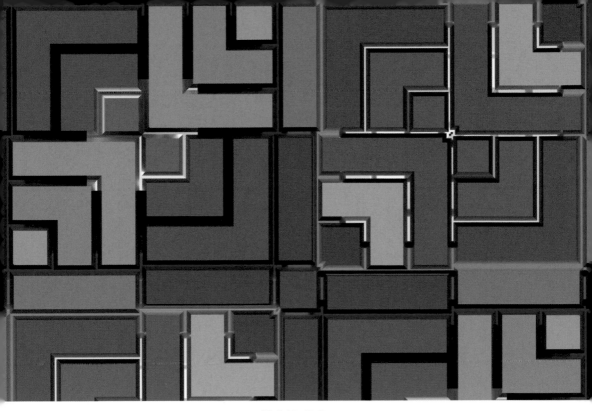

MAY 29

"Since Galileo's time, science has become steadily more mathematical. . . . It is virtually an article of faith for most theoreticians . . . that there exists a fundamental equation to describe the phenomenon they are studying. . . . Yet . . . it may eventually turn out that fundamental laws of nature do not need to be stated mathematically and that they are better expressed in other ways, like the rules governing the game of chess."

—GRAHAM FARMELO, FOREWORD TO *IT MUST BE BEAUTIFUL*, 2003

MAY 30

"I have photographed many people: artists, writers, and scientists, among others. In speaking about their work, mathematicians use the words 'elegance,' 'truth,' and 'beauty' more than everyone else combined."

—MARIANA COOK, *MATHEMATICIANS*, 2009

MAY 31

"Suppose that we think of the integers lined up like dominoes.
The inductive step tells us that they are close enough for each domino to knock
over the next one, the base case tells us that the first domino falls over, the conclusion
is that they all fall over. The fault in this analogy is that it takes time for each domino
to fall and so a domino which is a long way along the line won't fall over for
a long time. Mathematical implication is outside time."

—PETER J. ECCLES, *AN INTRODUCTION TO MATHEMATICAL REASONING*, 1998

JUNE 1

"Our minds arise from the functioning of our physical brains,
and the very precise physical laws that underlie that functioning are
grounded in the mathematics that requires our brains for its existence."

—ROGER PENROSE, "WHAT IS REALITY?," *NEW SCIENTIST*, 2006

JUNE 2

"A mathematician, like a painter or poet, is a maker of patterns. If his patterns are more permanent than theirs, it is because they are made with ideas."

—G. H. HARDY, *A MATHEMATICIAN'S APOLOGY*, 1941

JUNE 3

"The language of mathematics differs from that of everyday life, because it is essentially a rationally planned language. . . . The study of a man's social life has not yet brought forth a Linnaeus. . . . Curiously enough, people who are most sensible about the need for planning other social amenities in a reasonable way are often slow to see the need for creating a rational and international language."

—LANCELOT HOGBEN, *MATHEMATICS FOR THE MILLION*, 1937

JUNE 4

"The simple equations that generate the convoluted Mandelbrot
fractal have been called the wittiest remarks ever made."

—JOHN ALLEN PAULOS, *ONCE UPON A NUMBER*, 1998

JUNE 5

"The chess-board is the world; the pieces are the phenomena
of the universe; the rules of the game are what we call the laws of Nature.
The player on the other side is hidden from us. We know that his play is always
fair, and patient. But also we know, to our cost, that he never overlooks
a mistake, or makes the smallest allowance for ignorance."

—THOMAS HENRY HUXLEY, *LAY SERMONS, ADDRESSES, AND REVIEWS*, 1888

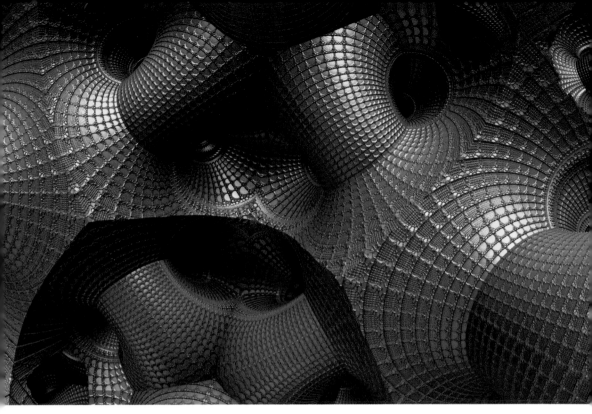

JUNE 6

"Mathematicians are only dealing with the structure of reasoning, and they do not really care what they are talking about. They do not even need to *know* what they are talking about. . . . But the physicist has meaning to all his phrases. . . . In physics, you have to have an understanding of the connection of words with the real world."

—RICHARD FEYNMAN, *THE CHARACTER OF PHYSICAL LAW*, 1965

JUNE 7

"The only reason I don't know more about love is because there just isn't more to know. In fact, I've reduced love to a mathematical formula: $\mathrm{Hdg}^k(X) = H^{2k}(X, \mathbf{Q}) \cap H^{k,k}(X)$. Actually, that's not right. That's the statement piece of the Hodge conjecture, but I'm sure you already knew that."

—JAROD KINTZ, *THE DAYS OF YAY ARE HERE!*, 2011

JUNE 8

"Hiding between all the ordinary numbers was an infinity of transcendental numbers whose presence you would never have guessed until you looked deeply into mathematics."

—CARL SAGAN, *CONTACT*, 1985

JUNE 9

BORN ON THIS DAY: John Edensor Littlewood, 1885

"The thing I want you especially to understand is this feeling of divine revelation. I feel that this structure was 'out there' all along I just couldn't see it. And now I can! This is really what keeps me in the math game—the chance that I might glimpse some kind of secret underlying truth, some sort of message from the gods."

—PAUL LOCKHART, *A MATHEMATICIAN'S LAMENT*, 2009

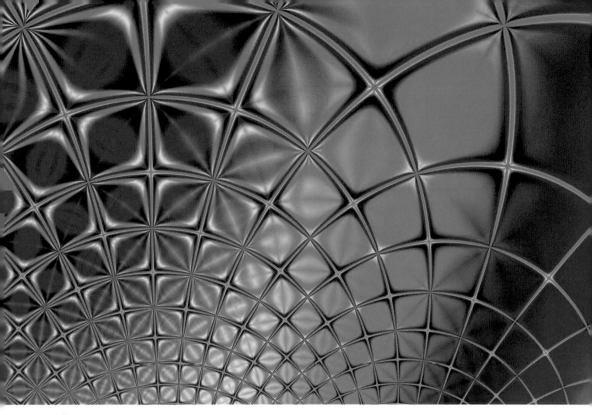

JUNE 10

"Until Einstein's time, scientists typically would observe things,
record them, then find a piece of mathematics that explained the results.
Einstein exactly reverses that process. He starts off with a beautiful piece of
mathematics that's based on some very deep insights into the way the universe
works and then, from that, makes predictions about what ought to happen in the
world. It's a stunning reversal to the usual ordering in which science is done. . . .
[Einstein demonstrated] the power of human creativity in the sciences . . ."

—SYLVESTER JAMES GATES, IN PETER TYSON'S "THE LEGACY OF $E = MC^2$," *NOVA*, 2005

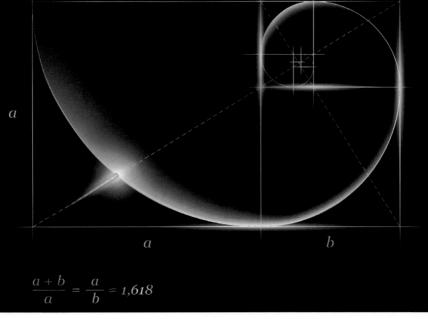

a

a

b

$$\frac{a+b}{a} = \frac{a}{b} = 1{,}618$$

JUNE 11

"The power of the Golden Section to create harmony arises from its unique capacity to unite the different parts of a whole so that each preserves its own identity, and yet blends into the greater pattern of a single whole. The Golden Section's ratio is an irrational, infinite number which can only be approximated. Yet such approximations are possible even within the limits of small whole numbers. This recognition filled the ancient Pythagoreans with awe: they sensed in it the secret power of cosmic order. It also led to the endeavors to realize the harmonies of such proportions in the patterns of daily life, thereby elevating life to an art."

—GYORGY DOCZI, *THE POWER OF LIMITS*, 1981

JUNE 12

"If people do not believe that mathematics is simple,
it is only because they do not realize how complicated life is."

—JOHN VON NEUMANN, ADDRESS AT ASSOCIATION FOR COMPUTING MACHINERY MEETING, 1947

JUNE 13

BORN ON THIS DAY: John Forbes Nash Jr., 1928; Grigori Perelman, 1966

"The pursuit of mathematics is a divine madness of the human spirit."

—ALFRED NORTH WHITEHEAD, *SCIENCE AND THE MODERN WORLD*, 1926

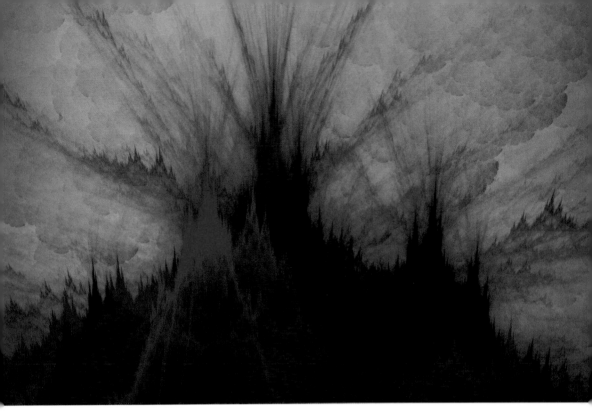

JUNE 14

BORN ON THIS DAY: Andrei Markov, 1856; Atle Selberg, 1917

"Most current mathematical research, since the 60s, is devoted to fancy situations: it brings solutions which nobody understands to questions nobody asked. Nevertheless, those who bring these solutions are called distinguished by the academic community. This word by itself gives a measure of the social distance: real life mathematics does not require distinguished mathematicians. On the contrary, it requires barbarians: people willing to fight, to conquer, to build, to understand, with no predetermined idea about which tool should be used."

—BERNARD BEAUZAMY, "REAL LIFE MATHEMATICS," *IRISH MATHEMATICS SOCIETY BULLETIN*, 2002

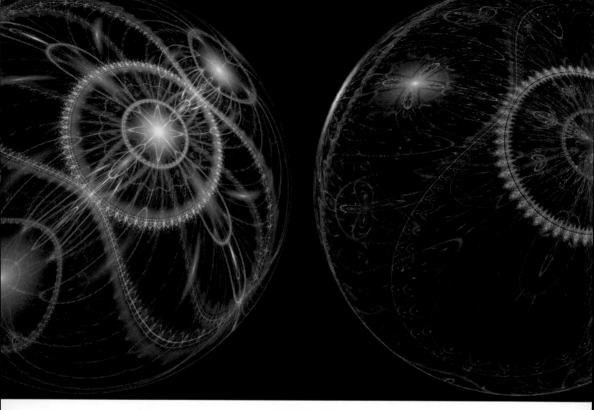

JUNE 15

"Every now and again one comes across an astounding result that closely relates two foreign objects which seem to have nothing in common. Who would suspect, for example, that on the average, the number of ways of expressing a positive integer n as a sum of two integral squares, $x^2 + y^2 = n$, is π."

—ROSS HONSBERGER, *MATHEMATICAL GEMS III*, 1985

JUNE 16

"Mathematical study and research are very suggestive of mountaineering. Whymper made several efforts before he climbed the Matterhorn in the 1860's and even then it cost the life of four of his party. Now, however, any tourist can be hauled up for a small cost, and perhaps does not appreciate the difficulty of the original ascent. So in mathematics, it may be found hard to realize the great initial difficulty of making a little step which now seems so natural and obvious, and it may not be surprising if such a step has been found and lost again."

—LOUIS JOEL MORDELL, *THREE LECTURES ON FERMAT'S LAST THEOREM*, 1921

JUNE 17

"For Catherine, time had lost its circadian rhythm; she had fallen into a tesseract of time, and day and night blended into one."

—SIDNEY SHELDON, *THE OTHER SIDE OF MIDNIGHT*, 1974

JUNE 18

"One of the big misapprehensions about mathematics that we perpetrate in our classrooms is that the teacher always seems to know the answer to any problem that is discussed. This gives students the idea that there is a book somewhere with all the right answers to all of the interesting questions, and that teachers know those answers. And if one could get hold of the book, one would have everything settled. That's so unlike the true nature of mathematics."

–LEON HENKIN, *TEACHING TEACHERS, TEACHING STUDENTS*, 1981

JUNE 19

BORN ON THIS DAY: Blaise Pascal, 1623

"Theorems are fun especially when you are the prover, but then the pleasure fades. What keeps us going are the unsolved problems."

—CARL POMERANCE, ADDRESS TO MATHEMATICAL ASSOCIATION OF AMERICA, 2000

JUNE 20

"The union of the mathematician with the poet, fervor with measure,
passion with correctness, this surely is the ideal."

—WILLIAM JAMES, *COLLECTED ESSAYS AND REVIEWS*, 1920

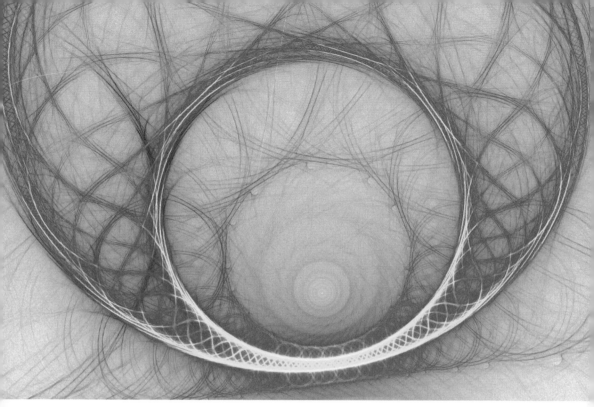

JUNE 21

BORN ON THIS DAY: Siméon Denis Poisson, 1781

"And what are these fluxions [derivatives]? The velocities of
evanescent increments. And what are these same evanescent increments?
They are neither finite quantities, nor quantities infinitely small, nor yet nothing.
May we not call them ghosts of departed quantities?"

—GEORGE BERKELEY, *THE ANALYST*, 1734

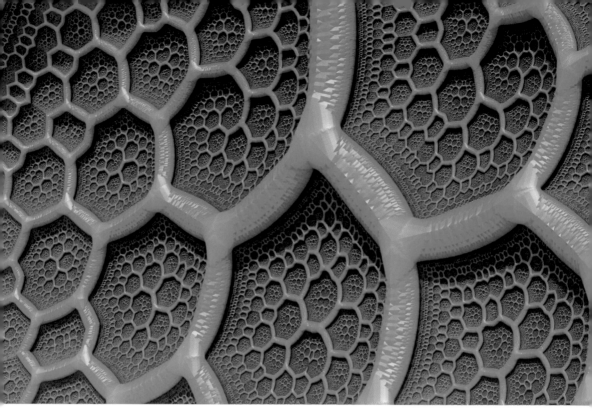

JUNE 22

BORN ON THIS DAY: Hermann Minkowski, 1864

"The pictures which science now draws of nature and which alone
seem capable of according with observational fact are mathematical pictures. . . .
From the intrinsic evidence of his creation, the Great Architect of the
Universe now begins to appear as a pure mathematician."

—JAMES HOPWOOD JEANS, *THE MYSTERIOUS UNIVERSE*, 1932

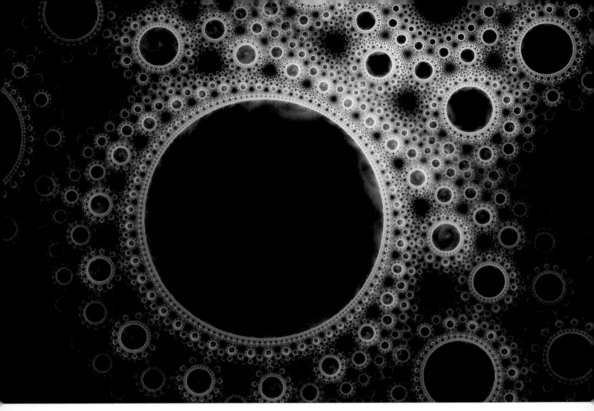

JUNE 23

BORN ON THIS DAY: Alan Mathison Turing, 1912

"We live on an island of knowledge surrounded by a sea of ignorance.
As our island of knowledge grows, so does the shore of our ignorance."

—JOHN A. WHEELER, *SCIENTIFIC AMERICAN*, 1992

JUNE 24

BORN ON THIS DAY: Oswald Veblen, 1880

"Geometry aims at knowledge of the eternal."

—PLATO, *REPUBLIC*, C. 380 BCE

JUNE 25

"$1/r^2$ has a nasty singularity at $r = 0$, but it did not
bother Newton—the Moon is far enough."

—EDWARD WITTEN, ADDRESS TO THE AMERICAN MATHEMATICAL SOCIETY, 1998

JUNE 26

"The problem is that we tend to live among the set of puny integers and generally ignore the vast infinitude of larger ones. How trite and limiting our view!"

—P. D. SCHUMER, *MATHEMATICAL JOURNEYS*, 2004

JUNE 27

"If you consider the width of a PC screen to be about a foot, and zoom into the M-set so that a piece of the complex plane 10^{-12} units wide fills the screen, then the M-set would extend to Jupiter. What are the chances, then, that in your fractal explorations you will find a piece of the Mandelbrot set never before seen with human eyes? Not only pretty good, but virtually certain. You may have heard of a company that for a fee will name a star after you and record it in a book. Maybe the same thing will soon be done with the Mandelbrot set!"

—TIM WEGNER AND MARK PETERSON, *FRACTAL CREATIONS*, 1991

JUNE 28

BORN ON THIS DAY: Henri Lebesgue, 1875

"Let no one ignorant of geometry enter here."

—INSCRIPTION ON PLATO'S ACADEMY, C. 380 BCE

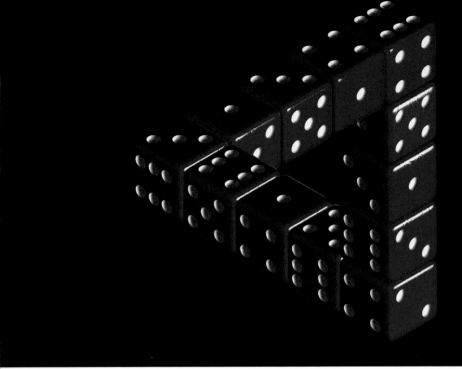

JUNE 29

"You get surreal numbers by playing games. I used to feel guilty in Cambridge that I spent all day playing games, while I was supposed to be doing mathematics. Then, when I discovered surreal numbers, I realized that playing games *is* math."

—JOHN H. CONWAY, PUBLIC LECTURE, PRINCETON UNIVERSITY, 1999

JUNE 30

"Who has not been amazed to learn that the function $y = e^x$, like a phoenix rising again from its own ashes, is its own derivative?"

—FRANÇOIS LE LIONNAIS, *GREAT CURRENTS OF MATHEMATICAL THOUGHT*, 1962

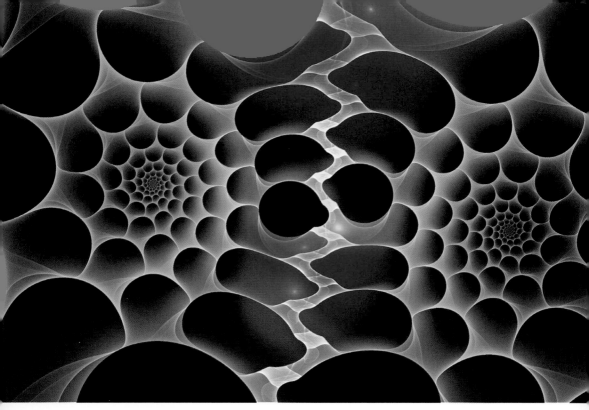

JULY 1

BORN ON THIS DAY: Gottfried Wilhelm Leibniz, 1646; Jean-Victor Poncelet, 1788

"The invention of the calculus was one of the great intellectual achievements of the 1600s. By one of those curious coincidences of mathematical history not one, but two men devised the idea—and almost simultaneously."

—DAVID M. BURTON, *THE HISTORY OF MATHEMATICS*, 2010

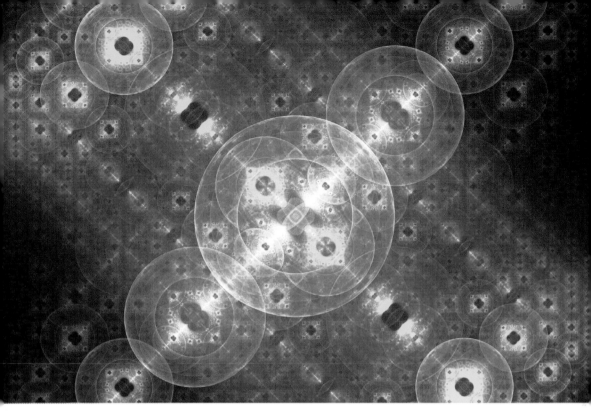

JULY 2

"Some of the men stood talking in this room, and at the right of the door
a little knot had formed round a small table, the center of which was the mathematics
student, who was eagerly talking. He had made the assertion that one could draw
through a given point more than one parallel to a straight line; Frau Hagenström had
cried out that this was impossible, and he had gone on to prove it so conclusively
that his hearers were constrained to behave as though they understood."

—THOMAS MANN, *LITTLE HERR FRIEDEMANN*, 1898

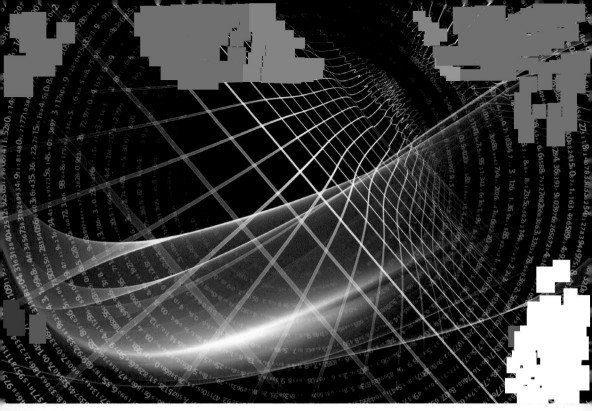

JULY 3

"Jeserac sat motionless within a whirlpool of numbers. He was fascinated by the way in which the numbers he was studying were scattered, apparently according to no laws, across the spectrum of integers."

—ARTHUR C. CLARKE, *THE CITY AND THE STARS*, 1956

JULY 4

"The computer provides a crucial jump in the way we do science
because it allows us to move in a realm where not all equations have solutions.
Nonlinear systems can display their feathers, and all of a sudden we see they
have a whole bunch of plumage that linear systems didn't have."

—NORMAN PACKARD, *OMNI*, 1992

JULY 5

"Mathematics has always shown a curious ability to be applicable to nature, and this may express a deep link between our minds and nature. We are the universe speaking out, a part of nature. So it is not surprising that our systems of logic and mathematics sing in tune with nature."

—GEORGE ZEBROWSKI, "LIFE IN GÖDEL'S UNIVERSE," *OMNI*, 1992

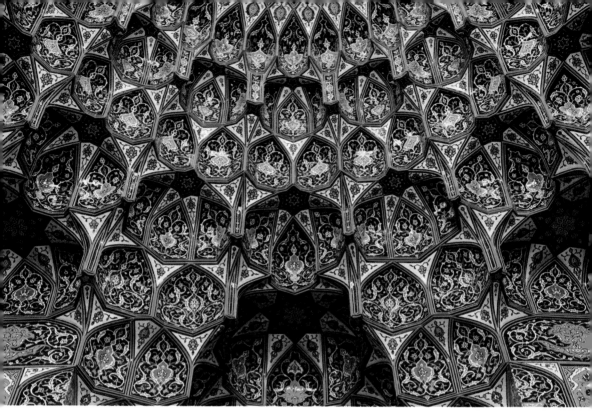

JULY 6

"Students must learn that mathematics is the most human of endeavors. Flesh and blood representatives of their own species engaged in a centuries-long creative struggle to uncover and to erect this magnificent edifice. And the struggle goes on today. On the very campuses where mathematics is presented and received as an inhuman discipline, cold and dead, new mathematics is created. As sure as the tides."

—J. D. PHILLIPS, "MATHEMATICS AS AN AESTHETIC DISCIPLINE," *HUMANISTIC MATHEMATICS NETWORK JOURNAL*, 1995

JULY 7

"The tantalizing and compelling pursuit of mathematical problems offers mental absorption, peace of mind amid endless challenges, repose in activity, battle without conflict, 'refuge from the goading urgency of contingent happenings,' and the sort of beauty changeless mountains present to senses tried by the present-day kaleidoscope of events."

—MORRIS KLINE, *MATHEMATICS IN WESTERN CULTURE*, 1953

JULY 8

"Simple shapes are inhuman. They fail to resonate with the way nature organizes itself or with the way human perception sees the world."

—JAMES GLEICK, *CHAOS*, 1987

JULY 9

"Gel'fand amazed me by talking of mathematics as though it were poetry. He once said about a long paper bristling with formulas that it contained the vague beginnings of an idea which he could only hint at and which he had never managed to bring out more clearly. I had always thought of mathematics as being much more straightforward: a formula is a formula, and an algebra is an algebra, but Gel'fand found hedgehogs lurking in the rows of his spectral sequences!"

—DUSA MCDUFF, *MATHEMATICAL NOTICES*, 1991

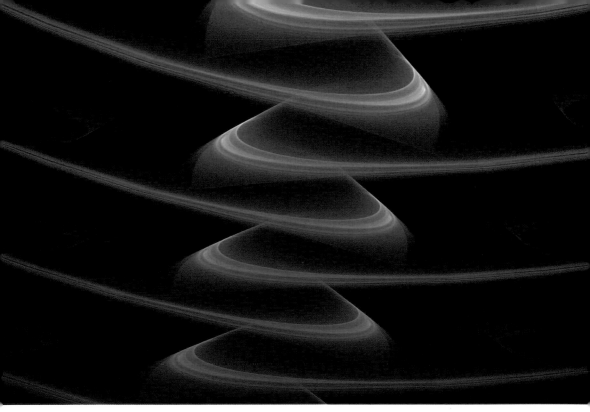

JULY 10

"Mathematical rigor is like clothing; in its style it ought to
suit the occasion, and it diminishes comfort and restrains freedom
of movement if it is either too loose or too tight."

—G. F. SIMMONS, *THE MATHEMATICAL INTELLIGENCER*, 1991

JULY 11

"There are two versions of math in the lives of many Americans: the strange and boring subject that they encountered in classrooms and an interesting set of ideas that is the math of the world, and is curiously different and surprisingly engaging. Our task is to introduce this second version to today's students, get them excited about math, and prepare them for the future."

—JO BOALER, *WHAT'S MATH GOT TO DO WITH IT?*, 2008

JULY 12

"For those, like me, who are not mathematicians, the computer can
be a powerful friend to the imagination. Like mathematics, it doesn't only
stretch the imagination. It also disciplines and controls it."

—RICHARD DAWKINS, *THE BLIND WATCHMAKER*, 1986

JULY 13

"To what purpose should People become fond of the Mathematicks and Natural Philosophy? . . . People very readily call Useless what they do not understand. It is a sort of Revenge. . . . One would think at first that if the Mathematicks were to be confin'd to what is useful in them, they ought only to be improv'd in those things which have an immediate and sensible Affinity with Arts, and the rest ought to be neglected as a Vain Theory. But this would be a very wrong Notion. As for Instance, the Art of Navigation hath a necessary Connection with Astronomy, and Astronomy can never be too much improv'd for the Benefit of Navigation. Astronomy cannot be without Optics by reason of Perspective Glasses: and both, as all parts of the Mathematicks are grounded upon Geometry. . . ."

—BERNARD LE BOVIER DE FONTENELLE, "OF THE USEFULNESS OF MATHEMATICAL LEARNING," 1699

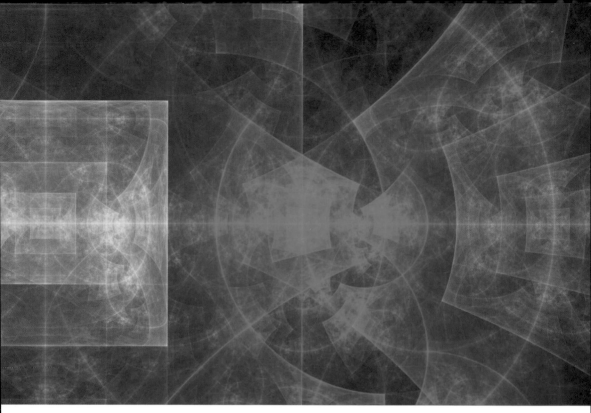

JULY 14

"Euclid alone has looked on Beauty bare.
Let all who prate of beauty hold their peace,
And lay them prone upon the earth and cease
To ponder on themselves, the while they stare
At nothing, intricately drawn nowhere
In shapes of shifting lineage; let geese
Gabble and hiss, but heroes seek release
From dusty bondage into luminous air."

—EDNA ST. VINCENT MILLAY, "EUCLID ALONE HAS LOOKED ON BEAUTY BARE," 1922

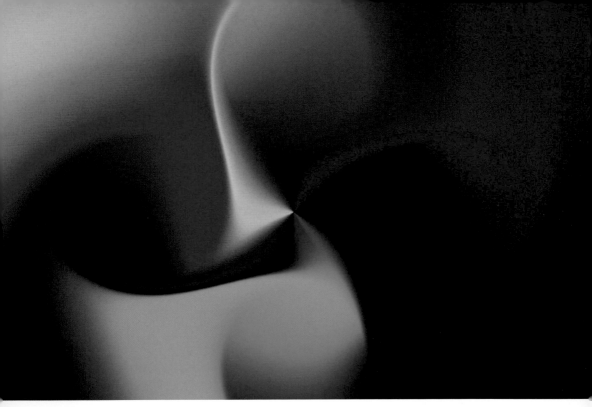

JULY 15

BORN ON THIS DAY: Stephen Smale, 1930

"There are in this world optimists who feel that any symbol that
starts off with an integral sign must necessarily denote something that will have
every property that they should like an integral to possess. This of course is quite
annoying to us rigorous mathematicians; what is even more annoying is that
by doing so they often come up with the right answer."

—E. J. MCSHANE, "INTEGRALS DEVISED FOR SPECIAL PURPOSES,"
BULLETIN OF THE AMERICAN MATHEMATICAL SOCIETY, 1963

JULY 16

BORN ON THIS DAY: Julius Plücker, 1801

"The curious inversion of Plattner's right and left sides is proof that he has moved out of our space into what is called the Fourth Dimension, and that he has returned again to our world."

—H. G. WELLS, "THE PLATTNER STORY," 1896

JULY 17

"Physicists have come to realize that mathematics, when used with sufficient care, is a proven pathway to truth."

—BRIAN GREENE, *THE FABRIC OF THE COSMOS*, 2005

JULY 18

"When we look back at the scientific revolution from our vantage point
of three centuries and attempt to understand the momentous transformation of
Western thought by isolating its central characteristic, the ever greater role of
mathematics and of quantitative modes of thought insistently catches our eye—what
Alexandre Koyré dubbed the geometrization of nature. Initiated in the sixteenth and
seventeenth centuries, the geometrization of nature has proceeded with gathering
momentum ever since. To be a scientist today is to understand and to do mathematics;
such is perhaps our most distinctive legacy from the scientific revolution."

—RICHARD S. WESTFALL, "NEWTON'S SCIENTIFIC PERSONALITY," *JOURNAL OF THE HISTORY OF IDEAS*, 1987

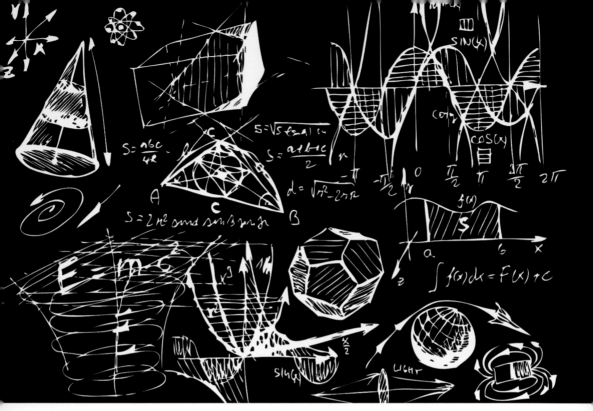

JULY 19

"This property of human languages—their resistance to algorithmic processing—is perhaps the ultimate reason why only mathematics can furnish an adequate language for physics. It is not that we lack words for expressing all this $E = mc^2$ and $\int e^{iS(\phi)}D\phi$ stuff . . . the point is that we still would not be able to do anything with these great discoveries if we had only words for them. . . . Miraculously . . . even very high level abstractions can somehow reflect reality: knowledge of the world discovered by physicists can be expressed only in the language of mathematics."

—YURI I. MANIN, "MATHEMATICAL KNOWLEDGE: INTERNAL, SOCIAL, AND CULTURAL ASPECTS,"
SELECTED ESSAYS OF YURI I. MANIN, 2007

JULY 20

"From Man or Angel the great Architect did wisely to conceal, and not divulge,
his secrets, to be scanned by them who ought rather admire. Or, if they list to try
conjecture, he his fabric of the Heavens hath left to their disputes—perhaps to move
His laughter at their quaint opinions wide. Hereafter, when they come to model
Heaven and calculate the stars: how they will wield the mighty frame: how build,
unbuild, contrive to save appearances; how gird the Sphere with Centric
and eccentric scribbled o'er, cycle and epicycle, orb in orb."

—JOHN MILTON, *PARADISE LOST,* 1667

JULY 21

"The main duty of the historian of mathematics, as well as his fondest privilege, is to explain the humanity of mathematics, to illustrate its greatness, beauty and dignity, and to describe how the incessant efforts and accumulated genius of many generations have built up that magnificent monument, the object of our most legitimate pride as men, and of our wonder, humility and thankfulness, as individuals. The study of the history of mathematics will not make better mathematicians but gentler ones, it will enrich their minds, mellow their hearts, and bring out their finer qualities."

—GEORGE SARTON, "THE STUDY OF THE HISTORY OF MATHEMATICS,"
AMERICAN MATHEMATICS MONTHLY, 1995

JULY 22

"The biologist can push it back to the original protist, and the chemist can push it back to the crystal, but none of them touch the real question of why or how the thing began at all. The astronomer goes back untold millions of years and ends in gas and emptiness, and then the mathematician sweeps the whole cosmos into unreality and leaves one with mind as the only thing of which we have any immediate apprehension. *Cogito ergo sum, ergo omnia esse videntur.* All this bother, and we are no further than Descartes. Have you noticed that the astronomers and mathematicians are much the most cheerful people of the lot? I suppose that perpetually contemplating things on so vast a scale makes them feel either that it doesn't matter a hoot anyway, or that anything so large and elaborate must have some sense in it somewhere."

—DOROTHY L. SAYERS AND R. EUSTACE, *THE DOCUMENTS IN THE CASE*, 1930

JULY 23

"By relieving the brain of all unnecessary work, a good notation
sets it free to concentrate on more advanced problems, and,
in effect, increases the mental power of the race."

—ALFRED NORTH WHITEHEAD, *AN INTRODUCTION TO MATHEMATICS*, 1911

JULY 24

"We may have very good reason for saying that we are ourselves beings of four dimensions and we are turned towards the third dimension with only one of our sides, i.e. with only a small part of our being. Only this part of us lives in three dimensions, and we are conscious only of this part as our body. The greater part of our being lives in the fourth dimension, but we are unconscious of this greater part of ourselves. Or it would be still more true to say that we live in a four-dimensional world, but are conscious of ourselves only in a three-dimensional world."

—P. D. OUSPENSKY, *THE FOURTH DIMENSION*, 1908

JULY 25

"Tyndall declared that he saw in Matter the promise and potency of all forms of life, and with his Irish graphic lucidity made a picture of a world of magnetic atoms, each atom with a positive and a negative pole, arranging itself by attraction and repulsion in orderly crystalline structure. Such a picture is dangerously fascinating to thinkers oppressed by the bloody disorders of the living world. Craving for purer subjects of thought, they find in the contemplation of crystals and magnets a happiness more dramatic and less childish than the happiness found by mathematicians in abstract numbers, because they see in the crystals beauty and movement without the corrupting appetites of fleshly vitality."

—GEORGE BERNARD SHAW, *BACK TO METHUSELAH*, 1921

JULY 26

"Nothing has afforded me so convincing a proof of the unity of the Deity as these purely mental conceptions of numerical and mathematical science which have been by slow degrees vouchsafed to man, and are still granted in these latter times by the Differential Calculus, now superseded by the Higher Algebra, all of which must have existed in that sublimely omniscient Mind from eternity."

—MARTHA SOMERVILLE, *PERSONAL RECOLLECTIONS, FROM EARLY LIFE TO OLD AGE, OF MARY SOMERVILLE*, 1874

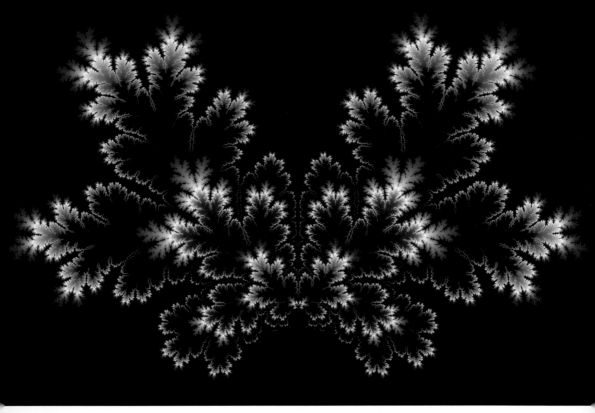

JULY 27

BORN ON THIS DAY: Johann Bernoulli, 1667; Ernst Zermelo, 1871

"The successes of the differential equation paradigm were impressive and extensive. Many problems, including basic and important ones, led to equations that could be solved. A process of self-selection set in, whereby equations that could not be solved were automatically of less interest than those that could."

—IAN STEWART, *DOES GOD PLAY DICE?*, 1989

JULY 28

"The mathematic, then, is an art. As such it has its styles and style periods.
It is not, as the layman and the philosopher (who is in this matter a layman too)
imagine, substantially unalterable, but subject like every art to unnoticed changes from
epoch to epoch. The development of the great arts ought never to be treated without
an (assuredly not unprofitable) side-glance at contemporary mathematics."

—OSWALD SPENGLER, "MEANING OF NUMBERS," IN JAMES NEWMAN'S
THE WORLD OF MATHEMATICS (VOL. 4), 1956

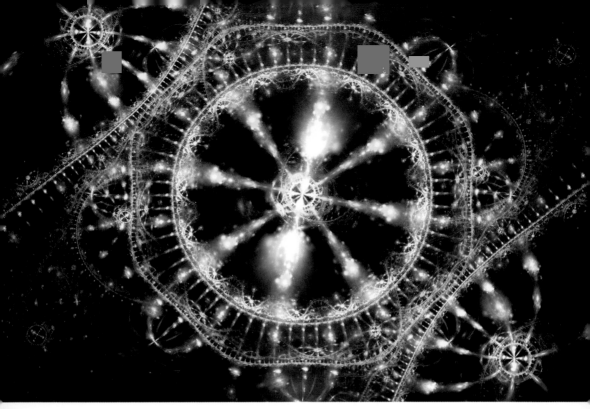

JULY 29

"The enormous usefulness of mathematics in natural sciences is something bordering on the mysterious, and there is no rational explanation for it. It is not at all natural that 'laws of nature' exist, much less that man is able to discover them. The miracle of the appropriateness of the language of mathematics for the formulation of the laws of physics is a wonderful gift which we neither understand nor deserve."

—EUGENE P. WIGNER, "THE UNREASONABLE EFFECTIVENESS OF MATHEMATICS IN THE NATURAL SCIENCES," *COMMUNICATIONS ON PURE AND APPLIED MATHEMATICS*, 1960

JULY 30

"Do not disturb my circles!"

—ARCHIMEDES, C. 212 BCE

"The pi machine prints the digits of pi in a surreal typeface where every digit is half as wide as its predecessor. The complete printout fits on an index card, but not even the most powerful electron microscope will reveal the last digit."

—WILLIAM POUNDSTONE, *LABYRINTHS OF REASON*, 1988

∞

AUGUST 1

"One of the nicest things about mathematics, or anything else
you might care to learn, is that many of the things which can never be,
often are. You see . . . it's very much like your trying to reach Infinity. You
know that it's there, but you just don't know where—but just because you
can never reach it doesn't mean that it's not worth looking for."

—NORTON JUSTER, *THE PHANTOM TOLLBOOTH*, 1988

AUGUST 2

"If we go back to our checker game, the fundamental laws are rules
by which the checkers move. Mathematics may be applied in the complex
situation to figure out what in given circumstances is a good move to make. But
very little mathematics is needed for the simple fundamental character of the
basic laws. They can be simply stated in English for checkers."

—RICHARD FEYNMAN, *THE CHARACTER OF PHYSICAL LAW*, 1965

AUGUST 3

"Contrary to popular belief, mathematics is a passionate subject. Mathematicians are driven by creative passions that are difficult to describe, but are no less forceful than those that compel a musician to compose or an artist to paint. The mathematician, the composer, the artist succumb to the same foibles as any human—love, hate, addictions, revenge, jealousies, desires for fame and money."

—THEONI PAPPAS, *MATHEMATICAL SCANDALS*, 1997

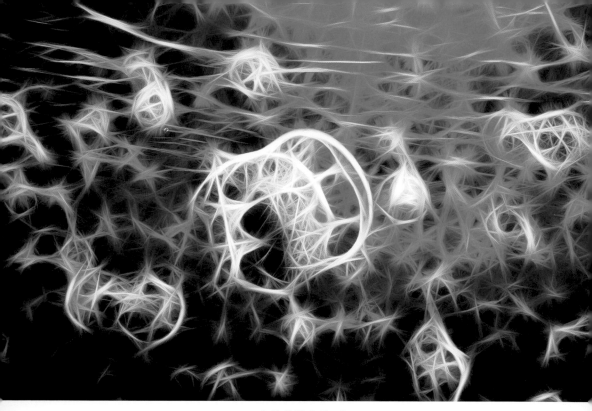

AUGUST 4

BORN ON THIS DAY: William Rowan Hamilton, 1805; John Venn, 1834

"Some men still believe in the mathematical design of nature. They may grant that many of the earlier mathematical theories of physical phenomena were imperfect, but they point to the continuing improvements that not only embrace more phenomena but offer far more accurate agreement with observations. Thus Newtonian mechanics replaces Aristotelian mechanics, and the theory of relativity improved on Newtonian mechanics. Does not this history imply that there is design and that man is approaching closer and closer to the truth?"

—MORRIS KLINE, *MATHEMATICS: THE LOSS OF CERTAINTY*, 1980

AUGUST 5

BORN ON THIS DAY: Niels Hendrik Abel, 1802

"Mathematics is a wonderful, mad subject, full of imagination, fantasy, and creativity that is not limited by the petty details of the physical world, but only by the strength of our inner light."

—GREGORY CHAITIN, "LESS PROOF, MORE TRUTH," *NEW SCIENTIST*, 2007

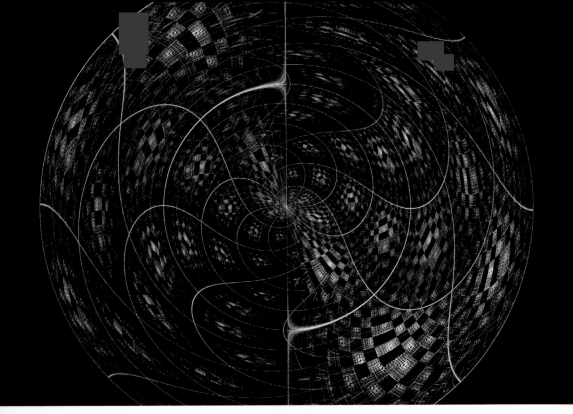

AUGUST 6

"Logic moves in one direction, the direction of clarity, coherence and structure. Ambiguity moves in the other direction, that of fluidity, openness, and release. Mathematics moves back and forth between these two poles . . . It is the interaction between these different aspects that gives mathematics its power."

—WILLIAM BYERS, *HOW MATHEMATICIANS THINK*, 2007

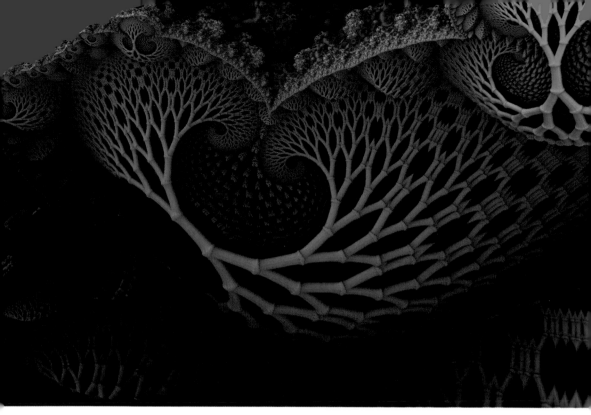

AUGUST 7

"The essential fact is simply that all the pictures which science now draws
of nature . . . are mathematical pictures. . . . It can hardly be disputed that nature
and our conscious mathematical minds work according to the same laws."

—JAMES HOPWOOD JEANS, *THE MYSTERIOUS UNIVERSE*, 1932

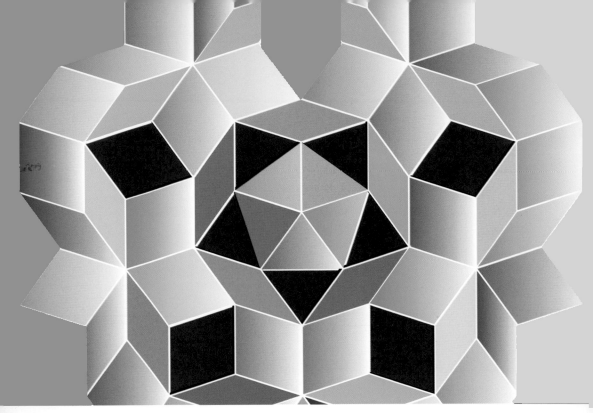

AUGUST 8

BORN ON THIS DAY: Roger Penrose, 1931

"The universe cannot be read until we have learnt the language and become familiar with the characters in which it is written. It is written in mathematical language, and the letters are triangles, circles and other geometrical figures, without which means it is humanly impossible to comprehend a single word."

—GALILEO GALILEI, *IL SAGGIATORE (THE ASSAYER)*, 1623

AUGUST 9

"Most of the papers which are submitted to the *Physical Review* are rejected, not because it is impossible to understand them, but because it is possible. Those which are impossible to understand are usually published."

—FREEMAN DYSON, "INNOVATION IN PHYSICS," *SCIENTIFIC AMERICAN*, 1958

AUGUST 10

"10th August, 1851: The night was chill [at the museum ball], and I dropped too suddenly from differential calculus into ladies' society, and I could not give myself freely to the change. After an hour's attempt so to do, I returned, cursing the mode of life I was pursuing. Next morning I had already shaken hands, however, with differential calculus and forgot the ladies. . . ."

—THOMAS ARCHER HIRST'S JOURNAL ENTRY FOR AUGUST 10, 1851

AUGUST 11

"The body of mathematics to which the calculus gives rise embodies
a certain swashbuckling style of thinking, at once bold and dramatic, given over
to large intellectual gestures and indifferent, in large measure, to any very detailed
description of the world. It is a style that has shaped the physical but not the biological
sciences, and its success in Newtonian mechanics, general relativity and
quantum mechanics is among the miracles of mankind."

—DAVID BERLINSKI, *A TOUR OF THE CALCULUS*, 1997

AUGUST 12

"It is known that there is an infinite number of worlds, simply because there is an infinite amount of space for them to be in. However, not every one of them is inhabited. Therefore, there must be a finite number of inhabited worlds. Any finite number divided by infinity is as near to nothing as makes no odds, so the average population of all planets in the Universe can be said to be zero. From this it follows that the population of the whole Universe is also zero, and that any people you may meet from time to time are merely the products of a deranged imagination."

—DOUGLAS ADAMS, *THE RESTAURANT AT THE END OF THE UNIVERSE*, 1980

AUGUST 13

"Mathematics, as much as music or any other art, is one of the means by which we rise to a complete self-consciousness. The significance of mathematics resides precisely in the fact that it is an art; by informing us of the nature of our own minds it informs us of much that depends on our minds."

—J.W.N. SULLIVAN, *ASPECTS OF SCIENCE*, 1925

AUGUST 14

BORN ON THIS DAY: Jean-Gaston Darboux, 1842

"God created the natural numbers, and all the rest is the work of man."

—LEOPOLD KRONECKER, ADDRESS AT THE *BERLINER NATURFORSCHER-VERSAMMLUNG*
(BERLIN NATURALIST MEETING), 1886

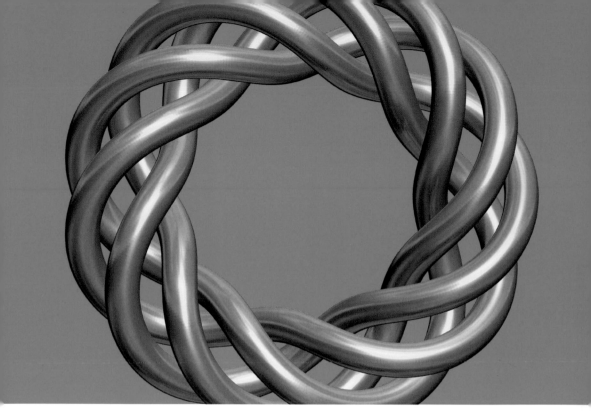

AUGUST 15

"We have a closed circle of consistency here: the laws of physics produce complex systems, and these complex systems lead to consciousness, which then produces mathematics, which can then encode in a succinct and inspiring way the very underlying laws of physics that give rise to it."

—PAUL DAVIES, *ARE WE ALONE?*, 1995

AUGUST 16

BORN ON THIS DAY: Arthur Cayley, 1821

"The mathematician is entirely free, within the limits of his imagination, to construct what worlds he pleases. What he is to imagine is a matter for his own caprice; he is not thereby discovering the fundamental principles of the universe nor becoming acquainted with the ideas of God. If he can find, in experience, sets of entities which obey the same logical scheme as his mathematical entities, then he has applied his mathematics to the external world; he has created a branch of science."

—J.W.N. SULLIVAN, *ASPECTS OF SCIENCE*, 1925

AUGUST 17

BORN ON THIS DAY: Pierre de Fermat, 1601

"There is no study in the world which brings into more harmonious action all the faculties of the mind than [mathematics], . . . or, like this, seems to raise them, by successive steps of initiation, to higher and higher states of conscious intellectual being. . . ."

—JAMES JOSEPH SYLVESTER, PRESIDENTIAL ADDRESS TO SECTION A OF THE BRITISH ASSOCIATION, 1869

233

AUGUST 18

"For twenty pages perhaps, he read slowly, carefully, dutifully, with pauses for self-examination and working out examples. Then, just as it was working up and the pauses should have been more scrupulous than ever, a kind of swoon and ecstasy would fall on him, and he read ravening on, sitting up till dawn to finish the book, as though it were a novel. After that his passion was stayed; the book went back to the Library and he was done with mathematics till the next bout. Not much remained with him after these orgies, but something remained: a sensation in the mind, a worshiping acknowledgment of something isolated and unassailable, or a remembered mental joy at the rightness of thoughts coming together to a conclusion, accurate thoughts, thoughts in just intonation, coming together like unaccompanied voices coming to a close."

—SYLVIA TOWNSEND WARNER, *MR. FORTUNE'S MAGGOT*, 1927

AUGUST 19

"It seems that nobody is indifferent to fractals. In fact, many view their first encounter with fractal geometry as a totally new experience from the viewpoints of aesthetics as well as science."

—BENOÎT B. MANDELBROT, QUOTED IN HEINZ-OTTO PEITGEN AND PETER RICHTER'S *THE BEAUTY OF FRACTALS*, 1986

AUGUST 20

"At some point Tengo noticed that returning to reality from the world of
a novel was not as devastating a blow as returning from the world of mathematics.
Why should that have been? After much deep thought, he reached a conclusion. No
matter how clear the relationships of things might become in the forest of story,
there was never a clear-cut solution. That was how it differed from math."

—HARUKI MURAKAMI, *1Q84*, 2011

AUGUST 21

BORN ON THIS DAY: Augustin-Louis Cauchy, 1789

"Every mathematician worthy of the name has experienced . . . the state of lucid exaltation in which one thought succeeds another as if miraculously . . . this feeling may last for hours at a time, even for days. Once you have experienced it, you are eager to repeat it but unable to do it at will, unless perhaps by dogged work . . ."

—ANDRÉ WEIL, *THE APPRENTICESHIP OF A MATHEMATICIAN*, 1991

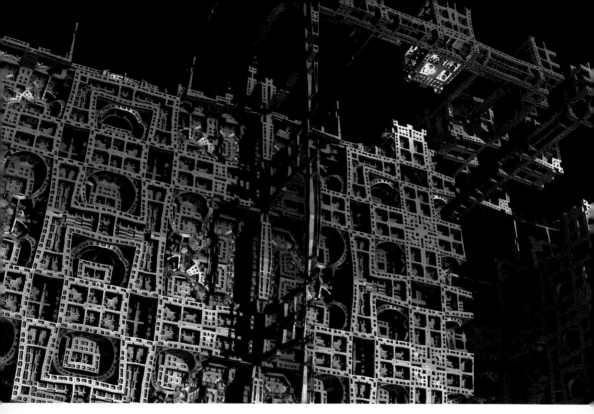

AUGUST 22

"Mathematics is the instrument by which the engineer tunnels our mountains, bridges our rivers, constructs our aqueducts, erects our factories, and makes them musical by the busy hum of spindles. Take away the results of the reasoning of mathematics, and there would go with it nearly all the material achievements which give convenience and glory to modern civilization."

—EDWARD BROOKS, *MENTAL SCIENCE AND CULTURE*, 1891

AUGUST 23

"Our story has a silent and immobile hero: the digital computer.
There can be little doubt that computers have acted as the most forceful
forceps in extracting fractals from the dark recesses of abstract mathematics
and delivering their geometrical intricacies into broad daylight."

—MANFRED SCHROEDER, *FRACTALS, CHAOS, POWER LAWS*, 1989

AUGUST 24

"Our federal income tax law defines the tax y to be paid in terms
of the income x; it does so in a clumsy enough way by pasting several linear
functions together, each valid in another interval or bracket of income. An archeologist
who, five thousand years from now, shall unearth some of our income tax returns
together with relics of engineering works and mathematical books, will probably
date them a couple of centuries earlier, certainly before Galileo and Vieta."

—HERMANN WEYL, "THE MATHEMATICAL WAY OF THINKING," *SCIENCE*, 1940

AUGUST 25

"Most people would have probably lost count around seven.
This was, Harry knew from his extensive reading on logic and arithmetic,
the largest number that most people could visually appreciate. Put seven dots
on a page, and most people can take a quick glance and declare,
'Seven.' Switch to eight, and the majority of humanity was lost."

—JULIA QUINN, *WHAT HAPPENS IN LONDON*, 2009

AUGUST 26

BORN ON THIS DAY: Johann Heinrich Lambert, 1728; Edward Witten, 1951

"The mathematics involved in string theory . . . in subtlety and sophistication . . . vastly exceeds previous uses of mathematics in physical theories. . . . String theory has led to a whole host of amazing results in mathematics in areas that seem far removed from physics. To many this indicates that string theory must be on the right track. . ."

—SIR MICHAEL ATIYAH, "PULLING THE STRINGS," *NATURE*, 2005

AUGUST 27

BORN ON THIS DAY: Giuseppe Peano, 1858

"I believe that scientific knowledge has fractal properties, that no matter how much we learn, whatever is left, however small it may seem, is just as infinitely complex as the whole was to start with. That, I think, is the secret of the Universe."

—ISAAC ASIMOV, *I ASIMOV*, 1995

AUGUST 28

"It is all about numbers. It is all about sequence. It's the mathematical
logic of being alive. If everything kept to its normal progression, we would
live with the sadness—cry and then walk—but what really breaks us
cleanest are the losses that happen out of order."

—AIMEE BENDER, *AN INVISIBLE SIGN OF MY OWN*, 2001

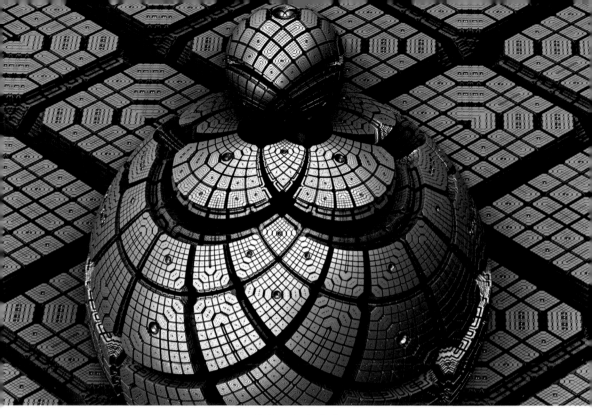

AUGUST 29

"An unspeakable horror seized me. There was a darkness; then a dizzy, sickening sensation of sight that was not like seeing; I saw a Line that was no Line; Space that was not Space; I was myself, and not myself. When I could find voice, I shrieked aloud in agony, 'Either this is madness or it is Hell.' 'It is neither,' calmly replied the voice of the Sphere, 'it is Knowledge; it is Three Dimensions; Open your eye once again and try to look steadily.'"

—EDWIN A. ABBOTT, *FLATLAND*, 1884

AUGUST 30

"Mathematics provides a framework for dealing precisely with notions
of 'what is.' Computation provides a framework for dealing
precisely with notions of 'how to.'"

—HAROLD ABELSON AND GERALD JAY SUSSMAN WITH JULIE SUSSMAN, PREFACE TO THE FIRST EDITION,
STRUCTURE AND INTERPRETATION OF COMPUTER PROGRAMS, 1996

AUGUST 31

"The universe we inhabit, and its operational principles, exist
independently of our observation or understanding; mathematical models
of the universe . . . are descriptive tools that exist only in our minds. Mathematics
is at root a formal description of orderliness, and since the universe is orderly
(at least on scales of space-time . . . which [we can] observe), it should come
as no surprise that the real world is well modeled mathematically."

—KEITH BACKMAN, "THE DANGER OF MATHEMATICAL MODELS," *SCIENCE*, 2006

SEPTEMBER 1

"The Mandelbrot Set is more than a mathematical plaything. It offers
one way of exploring the behavior of dynamical systems in which equations
express how some quantity changes over time or varies from place to place.
Such equations arise in calculations of the orbit of a planet, the flow
of heat in a liquid, and countless other situations."

—IVARS PETERSON, "BORDERING ON INFINITY: FOCUSING ON THE MANDELBROT SET'S
EXTRAORDINARY BOUNDARY," *SCIENCE NEWS*, 1991

SEPTEMBER 2

"The dreamer, a soldier in repose, applied the methods of algebra to the structures of geometry, bone-setting the measured land, expressing his system in terms of constants, variables, and position coordinates, all arranged in due time on the scheme of crossed lines forming squares of equal size."

—DON DELILLO, *RATNER'S STAR*, 1976

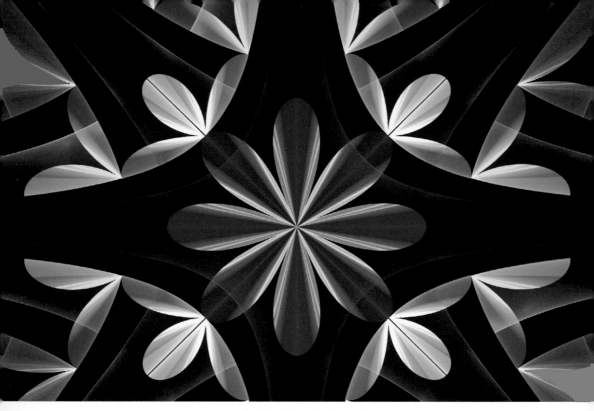

SEPTEMBER 3

BORN ON THIS DAY: James Joseph Sylvester, 1814

"The world of mathematics and physics, like the world of imagination,
is far removed from the tangible and visible; and yet, to the mathematician,
as to the poet, this world of pure form has an enduring reality."

—HELEN POLTZ, *IMAGINATION'S OTHER PLACE*, 1955

SEPTEMBER 4

"I'm one of those people who believe that life is a series of cycles—
wheels within wheels, some meshing with others, some spinning alone,
but all of them performing some finite, repeating function."

—STEPHEN KING, *FOUR PAST MIDNIGHT*, 1990

SEPTEMBER 5

"There can be no dull numbers, because if there were, the first
of them would be interesting on account of its dullness."

—MARTIN GARDNER, *FRACTAL MUSIC, HYPERCARDS AND MORE . . .* , 1991

SEPTEMBER 6

"We may in fact regard geometry as the most ancient branch of physics.
Without it I would have been unable to formulate the theory of relativity."

—ALBERT EINSTEIN, ADDRESS TITLED "GEOMETRY AND EXPERIENCE"
TO THE PRUSSIAN ACADEMY OF SCIENCES, 1921

SEPTEMBER 7

"Two of the most famous Baghdadi scholars, the philosopher Al-Kindi and the mathematician Al-Khawarizmi, were certainly the most influential in transmitting Hindu numerals to the Muslim world. Both wrote books on the subject during al-Ma'mun's reign, and it was their work that was translated into Latin and transmitted to the West, thus introducing Europeans to the decimal system, which was known in the Middle Ages only as Arabic numerals. But it would be many centuries before it was widely accepted in Europe. One reason for this was sociological: decimal numbers were considered for a long time as symbols of the evil Muslim foe."

—JIM AL-KHALILI, *PATHFINDERS*, 2010

SEPTEMBER 8

BORN ON THIS DAY: Marin Mersenne, 1588

"The Christians know that the mathematical principles, according to which the corporeal world was to be created, are coeternal with God. Geometry has supplied God with the models for the creation of the world. Within the image of God it has passed into man, and was certainly not received within through the eyes."

—JOHANNES KEPLER, *HARMONICE MUNDI (HARMONY OF THE WORLD)*, 1619

255

SEPTEMBER 9

"The Reader may here observe the Force of Numbers, which can be successfully applied, even to those things, which one would imagine are subject to no Rules. There are very few things which we know, which are not capable of being reduc'd to a Mathematical Reasoning; and when they cannot it's a sign our knowledge of them is very small and confus'd; and when a Mathematical Reasoning can be had it's as great a folly to make use of any other, as to grope for a thing in the dark, when you have a Candle standing by you."

—JOHN ARBUTHNOT, *OF THE LAWS OF CHANCE*, 1692

SEPTEMBER 10

"In the beginning, God said the four-dimensional divergence
of an antisymmetric, second rank tensor equals zero,
and there was Light, and it was good."

—MESSAGE ON T-SHIRT, AS REPORTED BY MICHIO KAKU IN "PARALLEL UNIVERSES,
THE MATRIX, AND SUPERINTELLIGENCE," KURZWEILAI.NET, 2003

SEPTEMBER 11

"Mathematics is a logical method . . . Mathematical propositions express no thoughts. In life it is never a mathematical proposition which we need, but we use mathematical propositions only in order to infer from propositions which do not belong to mathematics to others which equally do not belong to mathematics."

—LUDWIG WITTGENSTEIN, *TRACTATUS LOGICO PHILOSOPHICUS (LOGICAL-PHILOSOPHICAL TREATISE)*, 1922

SEPTEMBER 12

"Our minds are finite, and yet even in these circumstances of finitude we are surrounded by possibilities that are infinite, and the purpose of life is to grasp as much as we can out of that infinitude."

—ALFRED NORTH WHITEHEAD, *DIALOGUES*, 1954

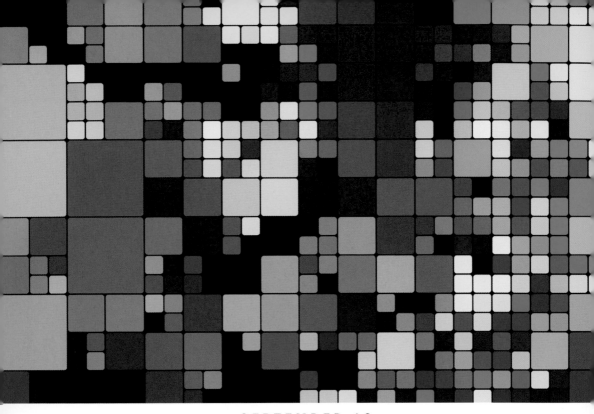

SEPTEMBER 13

"Prime numbers are what is left when you have taken all the patterns away.
I think prime numbers are like life. They are very logical but you could never work
out the rules, even if you spent all your time thinking about them."

—PAOLO GIORDANO, *THE SOLITUDE OF PRIME NUMBERS*, 2008

SEPTEMBER 14

"'Pray tell us, what's your favorite number?' . . .
'Shiva jumped up to the board, uninvited, and wrote 10,213,223' . . .
'And pray, why would this number interest us?'
'It is the only number that describes itself when you read it,
"One zero, two ones, three twos, two threes."'"

—ABRAHAM VERGHESE, *CUTTING FOR STONE*, 2010

SEPTEMBER 15

Jean-Pierre Serre, 1926

"One cannot escape the feeling that these mathematical formulae
have an independent existence and an intelligence of their own, that they
are wiser than we are, wiser even than their discoverers, that we
get more out of them than was originally put into them."

—HEINRICH HERTZ, QUOTED IN E. T. BELL'S *MEN OF MATHEMATICS*, 1937

SEPTEMBER 16

"I don't believe that math and nature respond to democracy.
Just because very clever people have rejected the role of the infinite,
their collective opinions, however weighty, won't persuade
mother nature to alter her ways. Nature is never wrong."

—JANNA LEVIN, *HOW THE UNIVERSE GOT ITS SPOTS*, 2003

SEPTEMBER 17

BORN ON THIS DAY: Bernhard Riemann, 1826

"Mathematics may be compared to a mill of exquisite workmanship, which grinds your stuff of any degree of fineness; but nevertheless, what you get out depends upon what you put in; and as the grandest mill in the world will not extract wheat-flour from peascods, so pages of formulae will not get a definite result out of loose data."

—THOMAS HENRY HUXLEY, *APHORISMS AND REFLECTIONS,* 1907

SEPTEMBER 18

BORN ON THIS DAY: Adrien Marie Legendre, 1752

"It is well known that geometry presupposes not only the concept of space but also the first fundamental notions for constructions in space as given in advance. It only gives nominal definitions for them, while the essential means of determining them appear in the form of axioms. The relationship of these presumptions is left in the dark; one sees neither whether and in how far their connection is necessary, nor a priori whether it is possible. From Euclid to Legendre, to name the most renowned of modern writers on geometry, this darkness has been lifted neither by the mathematicians nor the philosophers who have laboured upon it."

—BERNHARD RIEMANN, "ON THE HYPOTHESES WHICH LIE AT THE FOUNDATION OF GEOMETRY," 1854

SEPTEMBER 19

"I must study politics and war that my sons may have liberty to
study mathematics and philosophy. My sons ought to study mathematics and
philosophy, geography, natural history, naval architecture, navigation, commerce
and agriculture in order to give their children a right to study painting,
poetry, music, architecture, statuary, tapestry, and porcelain."

—JOHN ADAMS, LETTER TO ABIGAIL ADAMS, 1780

SEPTEMBER 20

"10^{50} is a long way from infinity."

—DANIEL SHANKS, *SOLVED AND UNSOLVED PROBLEMS IN NUMBER THEORY*, 2002

SEPTEMBER 21

"At every major step, physics has required, and frequently
stimulated, the introduction of new mathematical tools and concepts.
Our present understanding of the laws of physics, with their extreme
precision and universality, is only possible in mathematical terms."

—SIR MICHAEL ATIYAH, "PULLING THE STRINGS," *NATURE*, 2005

SEPTEMBER 22

"'Consider cotton prices,' Malcolm said. 'There are good records
of cotton prices going back more than a hundred years. When you study
fluctuations in cotton prices, you find that the graph of price fluctuations in the course
of a day looks basically like the graph for a week, which looks basically like the graph
for a year, or for ten years. . . . A day is like a whole life. You start out doing one
thing, but end up doing something else, plan to run an errand, but never get there. . . .
And at the end of your life, your whole existence has that same haphazard
quality, too. Your whole life has the same shape as a single day.'"

—MICHAEL CRICHTON, *JURASSIC PARK,* **1990**

SEPTEMBER 23

"Math is like water. It has a lot of difficult theories, of course, but its basic logic is very simple. Just as water flows from high to low over the shortest possible distance, figures can only flow in one direction. You just have to keep your eye on them for the route to reveal itself. That's all it takes. You don't have to do a thing. Just concentrate your attention and keep your eyes open, and the figures make everything clear to you. In this whole, wide world, the only thing that treats me so kindly is math."

—HARUKI MURAKAMI, *1Q84*, 2011

SEPTEMBER 24

BORN ON THIS DAY: Girolamo Cardano, 1501

"Art and science will eventually be seen to be as closely connected as arms to the body. Both are vital elements of order and its discovery. The word 'art' derives from the Indo-European base 'ar,' meaning to join or fit together. In this sense, science, in the attempt to learn how and why things fit, becomes art. And when art is seen as the ability to do, make, apply or portray in a way that withstands the test of time, its connection with science becomes more clear."

—SVEN CARLSON, *SCIENCE NEWS*, 1987

SEPTEMBER 25

"The belief that the underlying order of the world can
be expressed in mathematical form lies at the very heart of science.
So deep does this belief run that a branch of science is considered not to
be properly understood until it can be cast in mathematics."

—PAUL DAVIES, *THE MIND OF GOD*, 1992

SEPTEMBER 26

"I think scientists have a valid point when they bemoan the fact
that it's socially acceptable in our culture to be utterly ignorant of
math, whereas it is a shameful thing to be illiterate."

—JENNIFER OUELLETTE, *THE CALCULUS DIARIES*, 2010

SEPTEMBER 27

"Another mistaken notion connected with the law of large numbers is the idea that an event is more or less likely to occur because it has or has not happened recently. The idea that the odds of an event with a fixed probability increase or decrease depending on recent occurrences of the event is called the gambler's fallacy. For example, if Kerrich landed, say, 44 heads in the first 100 tosses, the coin would not develop a bias towards the tails in order to catch up! That's what is at the root of such ideas as 'Her luck has run out' and 'He is due.' That does not happen. For what it's worth, a good streak doesn't jinx you, and a bad one, unfortunately, does not mean better luck is in store."

—LEONARD MLODINOW, *THE DRUNKARD'S WALK*, 2008

SEPTEMBER 28

"This illustration from four dimensions . . . may lead us to vaster views of possible circumstances and existence; on the other hand it may teach us that the conception of such possibilities cannot, by any direct path, bring us closer to God. Mathematics may help us to measure and weigh the planets, to discover the materials of which they are composed, to extract light and warmth from the motion of water and to dominate the material universe; but even if by these means we could mount up to Mars or hold converse with the inhabitants of Jupiter or Saturn, we should be no nearer to the divine throne, except so far as these new experiences might develop in our modesty, respect for facts, a deeper reverence for order and harmony, and a mind more open to new observations and to fresh inferences from old truths."

—EDWIN A. ABBOTT, *THE SPIRIT ON THE WATERS*, 1897

SEPTEMBER 29

"He could not believe that any of them might actually hit somebody. If one did, what a nowhere way to go: killed by accident; slain not as an individual but by sheer statistical probability, by the calculated chance of searching fire, even as he himself might be at any moment. Mathematics! Mathematics! Algebra! Geometry!"

—JAMES JONES, *THE THIN RED LINE*, 1998

SEPTEMBER 30

"The mathematician's patterns, like the painter's or the poet's, must be beautiful; the ideas, like the colours or the words, must fit together in a harmonious way. Beauty is the first test: there is no permanent place in the world for ugly mathematics."

—G. H. HARDY, *A MATHEMATICIAN'S APOLOGY*, 1941

OCTOBER 1

"When I consider the small span of my life absorbed in the eternity of all time,
or the small part of space which I can touch or see engulfed by the infinite immensity
of spaces that I know not and that know me not, I am frightened and astonished. . . ."

—BLAISE PASCAL, *PENSÉES (THOUGHTS)*, 1669

OCTOBER 2

"We feel certain that the extraterrestrial message is a mathematical code
of some kind. Probably a number code. Mathematics is the one language we might
conceivably have in common with other forms of intelligent life in the universe.
As I understand it, there is no reality more independent of our perception
and more true to itself than mathematical reality."

—DON DELILLO, *RATNER'S STAR*, 1976

OCTOBER 3

"One might suppose that reality must be held to at all costs. However, though that may be the moral thing to do, it is not necessarily the most useful thing to do. The Greeks themselves chose the ideal over the real in their geometry and demonstrated very well that far more could be achieved by consideration of abstract line and form than by a study of the real lines and forms of the world; the greater understanding achieved through abstraction could be applied most usefully to the very reality that was ignored in the process of gaining knowledge."

—ISAAC ASIMOV, *UNDERSTANDING PHYSICS*, 1993

OCTOBER 4

"The world of ideas which it [mathematics] discloses or illuminates, the contemplation of divine beauty and order which it induces, the harmonious connexion of its parts, the infinite hierarchy and absolute evidence of the truths with which it is concerned, these, and such like, are the surest grounds of the title of mathematics to human regard, and would remain unimpeached and unimpaired were the plan of the universe unrolled like a map at our feet, and the mind of man qualified to take in the whole scheme of creation at a glance."

—JAMES JOSEPH SYLVESTER, PRESIDENTIAL ADDRESS TO SECTION A OF THE BRITISH ASSOCIATION, 1869

OCTOBER 5

"All beings sufficiently intelligent for interstellar communication must have a mathematics. . . . Yet their higher mathematics, their logic, their way of representing atomic structure, may differ radically from our own."

—WALTER SULLIVAN, *WE ARE NOT ALONE*, 1994

OCTOBER 6

BORN ON THIS DAY: Richard Dedekind, 1831; Robert Phelan Langlands, 1936

"From the outside, mathematics might look like one big lump. In fact, it is a huge subject that has many different subfields: algebra, number theory, analysis, geometry, and so on. In the world of mathematics, they look like disconnected continents. But the Langlands program connects different fields and, by doing so, tells us something about the unity of mathematics. It offers a glimpse of something beneath the surface that we don't understand."

—EDWARD FRENKEL, "A WORLD REVEALED" (INTERVIEW), *NEW SCIENTIST*, 2013

OCTOBER 7

"137.03 . . . is one of the greatest damn mysteries of physics: a magic number that comes to us with no understanding by man. You might say the 'hand of God' wrote that number, and 'we don't know how He pushed His pencil.'"

—RICHARD FEYNMAN, *QED*, 1985

OCTOBER 8

"I love mathematics not only for its technical applications, but principally because it is beautiful; because man has breathed his spirit of play into it, and because it has given him his greatest game—the encompassing of the infinite."

—ROZSA PETER, *PLAYING WITH INFINITY*, 1957

OCTOBER 9

"I turn away with fear and horror from this lamentable
plague of functions which do not have derivatives."

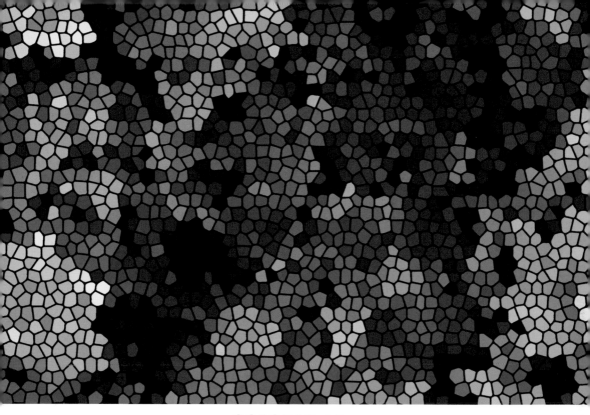

OCTOBER 10

"Mathematics alone make us feel the limits of our intelligence.
For we can always suppose in the case of an experiment that it is inexplicable
because we don't happen to have all the data. In mathematics we have all the data
and yet we don't understand. We always come back to the contemplation of our
human wretchedness. What force is in relation to our will, the impenetrable
opacity of mathematics is in relation to our intelligence."

—SIMONE WEIL, *NOTEBOOKS*, 1956

OCTOBER 11

"My work has always tried to unite the true with the beautiful and when
I had to choose one or the other I usually chose the beautiful."

—HERMANN WEYL, AS QUOTED BY FREEMAN DYSON IN WEYL'S OBITUARY IN *NATURE*, 1956

OCTOBER 12

"There may only be a small number of laws, which are self-consistent and which lead to complicated beings like ourselves. . . . And even if there is only one unique set of possible laws, it is only a set of equations. What is it that breathes fire into the equations and makes a universe for them to govern? Is the ultimate unified theory so compelling that it brings about its own existence?"

—STEPHEN HAWKING, *BLACK HOLES AND BABY UNIVERSES*, 1994

OCTOBER 13

"Most mathematicians prove many theorems in their lives, and the process whereby their name gets attached to one of them is very haphazard. For instance, Euler, Gauss, and Fermat each proved hundreds of theorems, many of them important ones, and yet their names are attached to just a few of them. Sometimes theorems acquire names that are incorrect. Most famously, perhaps, Fermat almost certainly did not prove 'Fermat's Last Theorem'; rather that name was attached by someone else, after his death, to a conjecture the French mathematician had scribbled in the margin of a textbook. And Pythagoras's theorem was known long before Pythagoras came onto the scene."

—KEITH DEVLIN, "NAMING THEOREMS," THE MATHEMATICAL ASSOCIATION OF AMERICA, 2005

OCTOBER 14

"To many, mathematics is a collection of theorems. For me,
mathematics is a collection of examples; a theorem is a statement
about a collection of examples and the purpose of proving
theorems is to classify and explain the examples . . ."

—JOHN B. CONWAY, *SUBNORMAL OPERATORS*, 1981

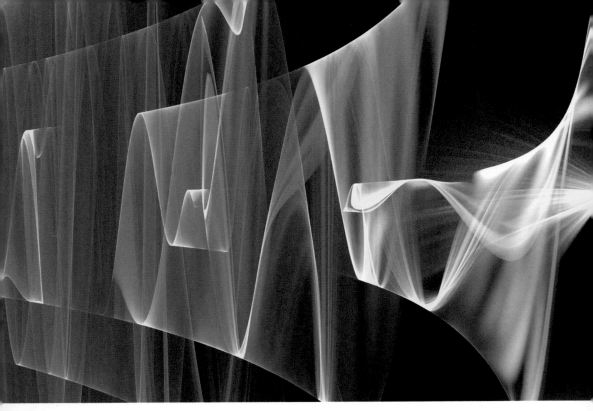

OCTOBER 15

"Some people can read a musical score and in their minds hear
the music. . . . Others can see, in their mind's eye, great beauty and structure
in certain mathematical functions. . . . Lesser folk, like me, need to hear music
played and see numbers rendered to appreciate their structures."

—PETER B. SCHROEDER, "PLOTTING THE MANDELBROT SET," *BYTE*, 1986

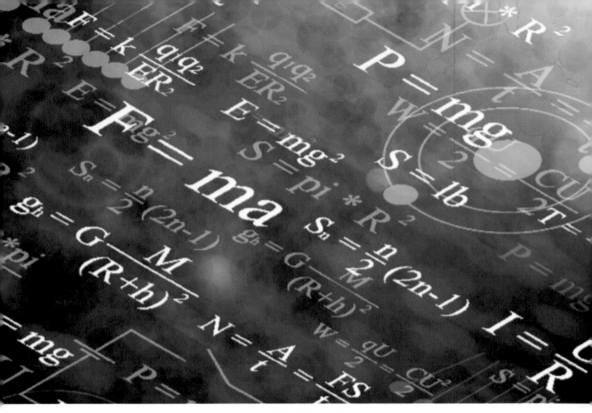

OCTOBER 16

"I often say that when you can measure what you are speaking about,
and express it in numbers, you know something about it; but when you cannot
express it in numbers, your knowledge is of a meager and unsatisfactory kind;
it may be the beginning of knowledge, but you have scarcely, in your thoughts,
advanced to the stage of science, whatever the matter may be."

—WILLIAM THOMSON, LECTURE ON ELECTRICAL UNITS OF MEASUREMENT, 1883

OCTOBER 17

"We could present spatially an atomic fact which contradicted the laws of physics, but not one which contradicted the laws of geometry."

—LUDWIG WITTGENSTEIN, *TRACTATUS LOGICO PHILOSOPHICUS* (*LOGICAL-PHILOSOPHICAL TREATISE*), 1922

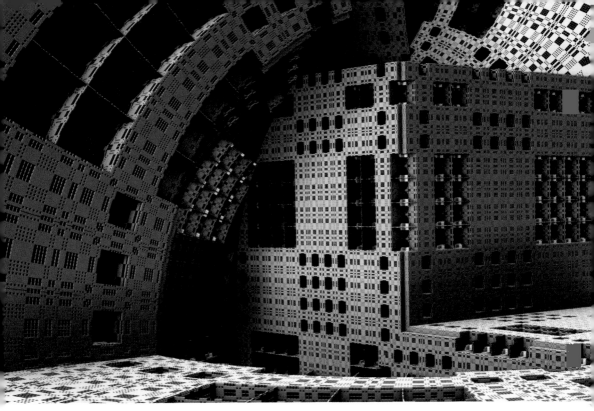

OCTOBER 18

"A man who devoted his life to it could perhaps succeed
in picturing to himself a fourth dimension."

—JULES HENRI POINCARÉ, "SPACE AND GEOMETRY," 1895

OCTOBER 19

"'In order to ascertain the height of the tree I must be in such a position
that the top of the tree is exactly in a line with the top of a measuring stick or
any straight object would do, such as an umbrella which I shall secure in an upright
position between my feet. Knowing then that the ratio that the height of the tree bears
to the length of the measuring stick must equal the ratio that the distance from my
eye to the base of the tree bears to my height, and knowing (or being able to find out)
my height, the length of the measuring stick and the distance from my eye
to the base of the tree, I can, therefore, calculate the height of the tree.'
'What is an umbrella?'"

—SYLVIA TOWNSEND WARNER, *MR. FORTUNE'S MAGGOT*, 1927

∞

OCTOBER 20

"Geometry is unique and eternal, and it shines in the mind of God.
The share of it which has been granted to man is one of
the reasons why he is the image of God."

—JOHANNES KEPLER, "CONVERSATION WITH THE SIDEREAL MESSENGER," 1610

OCTOBER 21

BORN ON THIS DAY: Martin Gardner, 1914

"The pure mathematician is much more of an artist than a scientist. He does not simply measure the world. He invents complex and playful patterns without the least regard for their practical applicability."

—ALAN WATTS, *BEYOND THEOLOGY*, 1964

OCTOBER 22

"The movement of humanity, arising as it does from innumerable
human wills, is continuous. To understand the laws of this continuous
movement is the aim of history. Only by taking an infinitesimally small unit
for observation (the differential of history, that is, the individual tendencies
of man) and attaining to the art of integrating them (that is, finding the sum
of these infinitesimals) can we hope to arrive at the laws of history."

—COUNT LEV NIKOLAYEVICH TOLSTOY, *WAR AND PEACE*, 1869

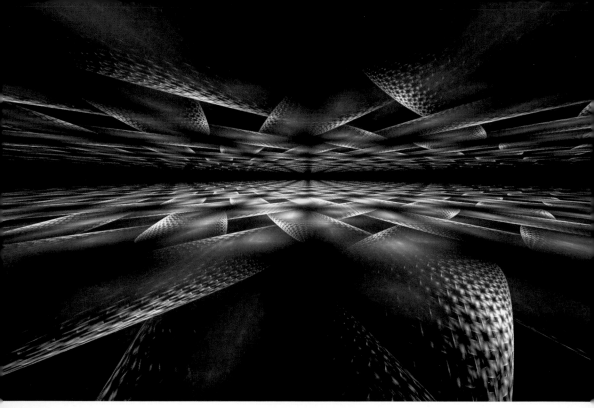

OCTOBER 23

"I believe that mathematical reality lies outside us, that our function is to discover or observe it, and that the theorems which we prove, and which we describe grandiloquently as our 'creations,' are simply the notes of our observations."

—G. H. HARDY, *A MATHEMATICIAN'S APOLOGY*, 1941

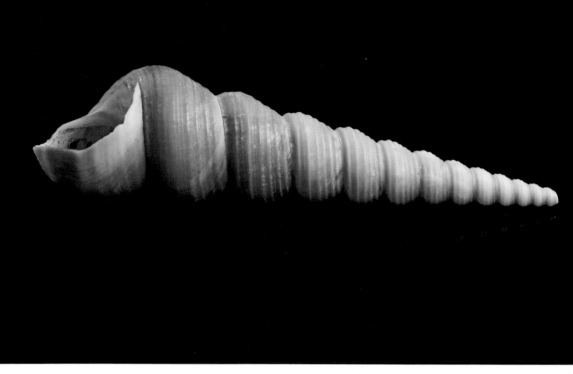

OCTOBER 24

"Cell and tissue, shell and bone, leaf and flower, are so many portions
of matter, and it is in obedience to the laws of physics that their particles have
been moved, moulded and conformed. They are no exceptions to the rule that God
always geometrizes. Their problems of form are in the first instance mathematical
problems, their problems of growth are essentially physical problems, and the
morphologist is, *ipso facto*, a student of physical science."

—D'ARCY WENTWORTH THOMPSON, *ON GROWTH AND FORM*, 1917

OCTOBER 25

BORN ON THIS DAY: Évariste Galois, 1811

"The essence of mathematics resides in its freedom."

—GEORG CANTOR, "ÜBER UNENDLICHE, LINEARE PUNKTMANNICHFALTIGKEITEN"
("ABOUT INFINITE, LINEAR MANIFOLDS OF POINTS"),
MATHEMATISCHE ANNALEN, 1883

OCTOBER 26

BORN ON THIS DAY: Ferdinand Georg Frobenius, 1849; Shiing-Shen Chern, 1911

"You can think of mathematics as the grand project of building this enormous jigsaw puzzle, with different groups of people working on different parts. Then, every once in a while, somebody finds a bridge between two parts, a way to assemble pieces together so that big chunks of the puzzle connect."

—EDWARD FRENKEL, "A WORLD REVEALED" (INTERVIEW), *NEW SCIENTIST*, 2013

OCTOBER 27

"But when great and ingenious artists behold their so inept performances, not undeservedly do they ridicule the blindness of such men; since sane judgment abhors nothing so much as a picture perpetrated with no technical knowledge, although with plenty of care and diligence. Now the sole reason why painters of this sort are not aware of their own error is that they have not learnt Geometry, without which no one can either be or become an absolute artist; but the blame for this should be laid upon their masters, who are themselves ignorant of this art."

—ALBRECHT DÜRER, *THE ART OF MEASUREMENT*, 1525

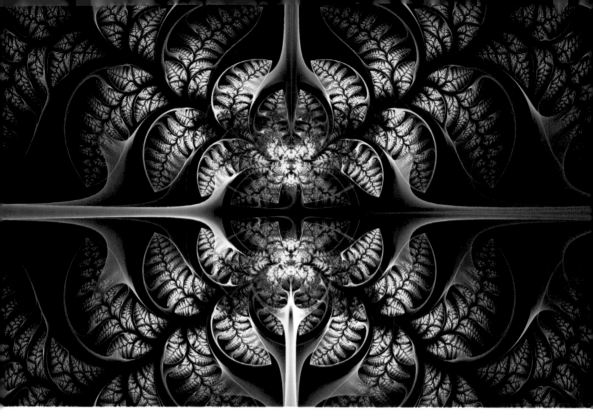

OCTOBER 28

"God has a transfinite book with all the theorems and their best proofs.
You don't really have to believe in God as long as you believe in the book."

—PAUL ERDŐS, QUOTED IN BRUCE SCHECHTER'S *MY BRAIN IS OPEN*, 1998

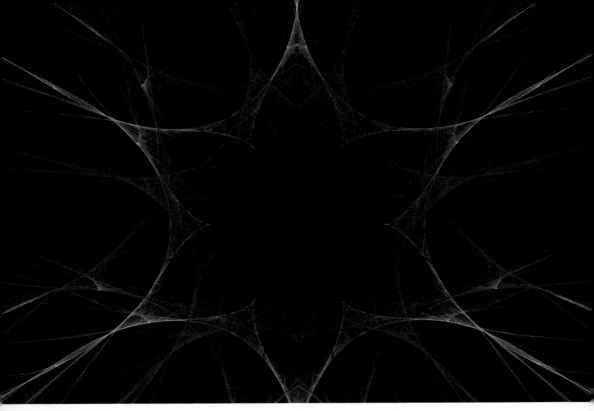

OCTOBER 29

"Whoever . . . proves his point and demonstrates the prime truth geometrically should be believed by all the world, for there we are captured."

—ALBRECHT DÜRER, *VIER BÜCHER VON MENSCHLICHER PROPORTION*
(*FOUR BOOKS ON HUMAN PROPORTION*), 1528

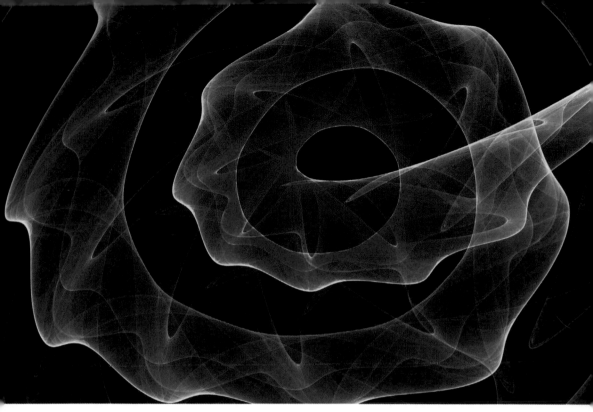

OCTOBER 30

BORN ON THIS DAY: William Thurston, 1946

"The greatest advantage to be derived from the study of geometry of more than three dimensions is a real understanding of the great science of geometry. Our plane and solid geometries are but the beginning of this science. The four-dimensional geometry is far more extensive than the three-dimensional, and all the higher geometries are more extensive than the lower."

—HENRY PARKER MANNING, *GEOMETRY OF FOUR DIMENSIONS*, 1914

OCTOBER 31

BORN ON THIS DAY: Karl Weierstrass, 1815

"In many cases, mathematics is an escape from reality.
The mathematician finds his own monastic niche and happiness
in pursuits that are disconnected from external affairs. Some practice
it as if using a drug. Chess sometimes plays a similar role. In their
unhappiness over the events of this world, some immerse themselves
in a kind of self-sufficiency in mathematics. (Some have
engaged in it for this reason alone.)"

—STANISLAW ULAM, *ADVENTURES OF A MATHEMATICIAN*, 1976

NOVEMBER 1

"The external world exists; the structure of the world is ordered; we know little about the nature of the order, nothing at all about why it should exist."

—MARTIN GARDNER, *ORDER AND SURPRISE*, 1950

NOVEMBER 2

BORN ON THIS DAY: George Boole, 1815

"There is what may perhaps be called the method of optimism which leads
us either willfully or instinctively to shut our eyes to the possibility of evil. Thus the
optimist who treats a problem in algebra or analytic geometry will say, if he stops to
reflect on what he is doing: 'I know that I have no right to divide by zero; but there
are so many other values which the expression by which I am dividing might have that
I will assume that the Evil One has not thrown a zero in my denominator this time.'"

—MAXIME BÔCHER, "THE FUNDAMENTAL CONCEPTIONS AND METHODS OF MATHEMATICS,"
BULLETIN OF THE AMERICAN MATHEMATICAL SOCIETY, 1904

NOVEMBER 3

"His way had therefore come full circle, or rather had taken the form
of an ellipse or a spiral, following as ever no straight unbroken line, for the
rectilinear belongs only to Geometry and not to Nature and Life."

—ISAAC ASIMOV, *SECOND FOUNDATION*, 1991

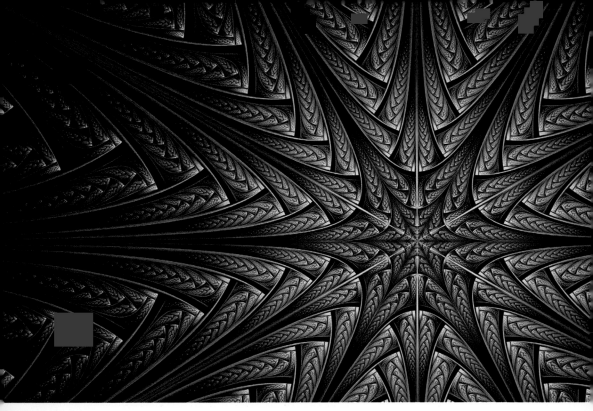

NOVEMBER 4

"Calculus need not be made easy; it is easy already. However, it is a subject which cannot be mastered by sheer memory. On the contrary, the student must accustom himself to memorizing as few formulas as possible and to reasoning out the correct attack in any situation."

—G. M. PETERSEN AND R. F. GRAESSER, *DIFFERENTIAL AND INTEGRAL CALCULUS*, 1961

NOVEMBER 5

"One can argue that mathematics is a human activity deeply
rooted in reality, and permanently returning to reality. From counting on
one's fingers to moon-landing to Google, we are doing mathematics in order to
understand, create, and handle things, and perhaps this understanding is mathematics
rather than intangible murmur of accompanying abstractions. Mathematicians are thus
more or less responsible actors of human history, like Archimedes helping to defend
Syracuse (and to save a local tyrant), Alan Turing cryptanalyzing Marshal Rommel's
intercepted military dispatches to Berlin, or John von Neumann suggesting
high altitude detonation as an efficient tactic of bombing."

—YURI I. MANIN, "MATHEMATICAL KNOWLEDGE: INTERNAL, SOCIAL, AND CULTURAL ASPECTS,"
MATHEMATICS AS METAPHOR: SELECTED ESSAYS, 2007

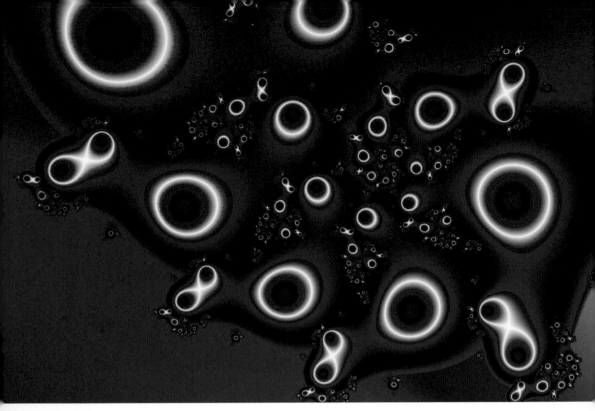

NOVEMBER 6

"Probability does pervade the universe—and in this sense, the old chestnut about baseball imitating life really has validity. The statistics of streaks and slumps, properly understood, do teach an important lesson about epistemology, and life in general. The history of a species, or any natural phenomenon, that requires unbroken continuity in a world of trouble, works like a batting streak. All are games of a gambler playing with a limited stake against a house with infinite resources. The gambler must eventually go bust. His aim can only be to stick around as long as possible, to have some fun while he's at it, and, if he happens to be a moral agent as well, to worry about staying the course with honor."

—STEPHEN JAY GOULD, "THE STREAK OF STREAKS," *THE NEW YORK REVIEW OF BOOKS*, 1988

NOVEMBER 7

"Steinhaus, with his predilection for metaphors, used to quote a Polish proverb, '*Forturny kolem sie tocza*' ['Luck runs in circles'], to explain why π, so intimately connected with circles, keeps cropping up in probability theory and statistics, the two disciplines which deal with randomness and luck."

—MARK KAC, *ENIGMAS OF CHANCE*, 1987

NOVEMBER 8

BORN ON THIS DAY: Gottlob Frege, 1848; Felix Hausdorff, 1868

"Fractal geometry will make you see everything differently. There is a danger in reading further. You risk the loss of your childhood vision of clouds, forests, flowers, galaxies, leaves, feathers, rocks, mountains, torrents of water, carpet, bricks, and much else besides. Never again will your interpretation of these things be quite the same."

—MICHAEL F. BARNSLEY, *FRACTALS EVERYWHERE*, 2000

NOVEMBER 9

BORN ON THIS DAY: Hermann Weyl, 1885

"In heterotic string theory . . . the right-handed bosons (carrier particles) go counterclockwise around the loop, their vibrations penetrating 22 compacted dimensions. The bosons live in a space of 26 dimensions (including time) of which 6 are the compacted 'real' dimensions, 4 are the dimensions of ordinary space-time, and the other 16 are deemed 'interior spaces'— mathematical artifacts to make everything work out right."

—MARTIN GARDNER, *THE NEW AMBIDEXTROUS UNIVERSE*, 1990

NOVEMBER 10

"We are not very pleased when we are forced to accept a mathematical truth by virtue of a complicated chain of formal conclusions and computations, which we traverse blindly, link by link, feeling our way by touch. We want first an overview of the aim and of the road; we want to understand the idea of the proof, the deeper context."

—HERMANN WEYL, *UNTERRICHTSBLÄTTER FÜR MATHEMATIK UND NATURWISSENSCHAFTEN* (*INSTRUCTION SHEETS FOR MATH AND SCIENCE*), 1932

NOVEMBER 11

"Armies of thinkers have been defeated by the enigma of why most fundamental laws of nature can be written down so conveniently as equations. Why is it that so many laws can be expressed as an absolute imperative, that two apparently unrelated quantities (the equation's right and left sides) are exactly equal? Nor is it clear why fundamental laws exist at all."

—GRAHAM FARMELO, FOREWORD TO HIS BOOK *IT MUST BE BEAUTIFUL*, 2003

NOVEMBER 12

"Numbers still gave Astrid pleasure. That was the great thing about numbers: it required no faith to believe that two plus two equaled four. And math never, ever condemned you for your thoughts and desires."

—MICHAEL GRANT, *FEAR*, 2013

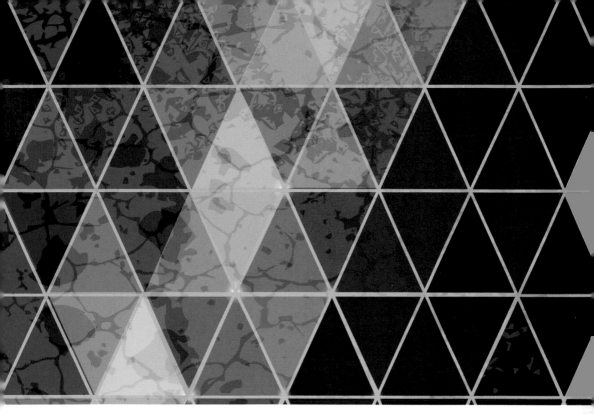

NOVEMBER 13

"As to the need of improvement there can be no question whilst the reign of Euclid continues. My own idea of a useful course is to begin with arithmetic, and then not Euclid but algebra. Next, not Euclid, but practical geometry. . . . Then not Euclid, but elementary vectors, conjoined with algebra, and applied to geometry. . . . Euclid might be an extra course for learned men, like Homer. But Euclid for children is barbarous."

—OLIVER HEAVISIDE, *ELECTRO-MAGNETIC THEORY*, 1893

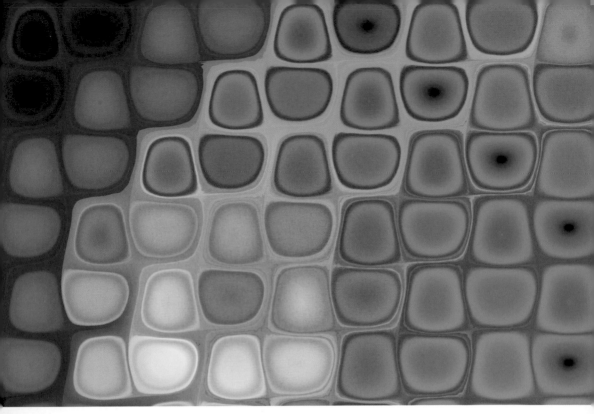

NOVEMBER 14

"Prime numbers are divisible only by 1 and by themselves. They hold their place in the infinite series of natural numbers, squashed, like all numbers, between two others, but one step further than the rest. They are suspicious, solitary numbers, which is why Mattia thought they were wonderful. Sometimes he thought that they had ended up in that sequence by mistake, that they'd been trapped, like pearls strung on a necklace. Other times he suspected that they too would have preferred to be like all others, just ordinary numbers, but for some reason they couldn't do it. This second thought struck him mostly at night, in the chaotic interweaving of images that comes before sleep, when the mind is too weak to tell itself lies."

—PAOLO GIORDANO, *THE SOLITUDE OF PRIME NUMBERS*, 2008

NOVEMBER 15

"The poetry of science is in some sense embodied in its great equations . . .
these equations can also be peeled. But their layers represent their attributes
and consequences, not their meanings. It is perfectly possible to imagine a universe
in which mathematical equations have nothing to do with the workings
of nature. Yet the marvelous thing is that they do."

—GRAHAM FARMELO, *IT MUST BE BEAUTIFUL*, 2003

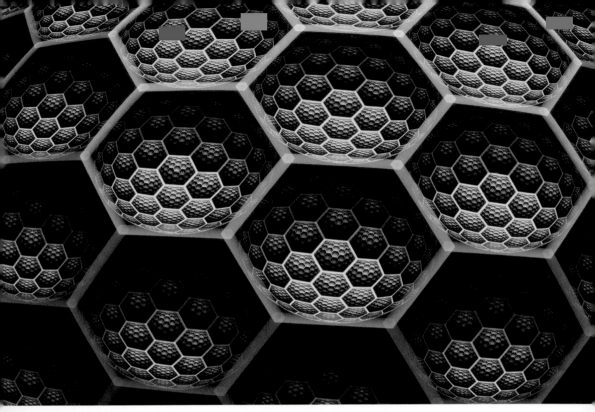

NOVEMBER 16

BORN ON THIS DAY: Jean le Rond d'Alembert, 1717

"In fact, Gentlemen, no geometry without arithmetic, no mechanics without geometry. . . . You cannot count upon success, if your mind is not sufficiently exercised on the forms and demonstrations of geometry, on the theories and calculations of arithmetic. . . . In a word, the theory of proportions is for industrial teaching, what algebra is for the most elevated mathematical teaching."

—JEAN-VICTOR PONCELET, *L'OUVERTURE DU COURS DE MÉCANIQUE INDUSTRIELLE DE METZ (INTRODUCTION TO THE COURSE IN INDUSTRIAL MECHANICS)*, 1827

NOVEMBER 17

BORN ON THIS DAY: August Ferdinand Möbius, 1790

"It hath been an old remark, that Geometry is an excellent Logic.
And it must be owned that when the definitions are clear; when the postulata
cannot be refused, nor the axioms denied; when from the distinct contemplation and
comparison of figures, their properties are derived, by a perpetual well-connected
chain of consequences, the objects being still kept in view, and the attention ever fixed
upon them; there is acquired a habit of reasoning, close and exact and methodical;
which habit strengthens and sharpens the mind, and being transferred to other
subjects is of general use in the inquiry after truth."

—GEORGE BERKELEY, *THE ANALYST*, 1898

NOVEMBER 18

"Perhaps to the student there is no part of elementary mathematics so repulsive as is spherical geometry."

—PETER GUTHRIE TAIT, "QUATERNIONS," *ENCYCLOPAEDIA BRITANNICA*, 1911

NOVEMBER 19

"Geometry, which should only obey Physics, when united with it sometimes commands it. If it happens that the question which we wish to examine is too complicated for all the elements to be able to enter into the analytical comparison we wish to make, we separate the more inconvenient [elements], we substitute others for them, less troublesome but also less real, and we are surprised to arrive, notwithstanding a painful labour, only at a result contradicted by nature; as if after having disguised it, cut it short or altered it, a purely mechanical combination could give it back to us."

—JEAN LE ROND D'ALEMBERT, *ESSAI D'UNE NOUVELLE THÉORIE DE LA RÉSISTANCE DES FLUIDS*
(*ESSAY ON A NEW THEORY OF FLUID RESISTANCE*), 1752

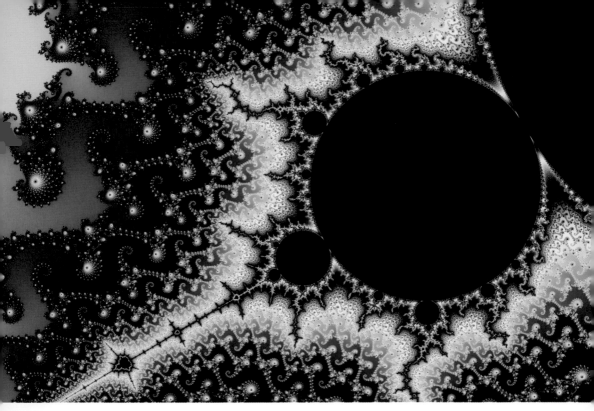

NOVEMBER 20

BORN ON THIS DAY: Benoît B. Mandelbrot, 1924

"It would seem that the Mandelbrot set is not just part of our minds, but it has a reality of its own . . . The computer is being used in essentially the same way that an experimental physicist uses a piece of experimental apparatus to explore the structure of the physical world. The Mandelbrot set is not an invention of the human mind: it was a discovery. Like Mount Everest, the Mandelbrot set is just there."

—ROGER PENROSE, *THE EMPEROR'S NEW MIND*, 1999

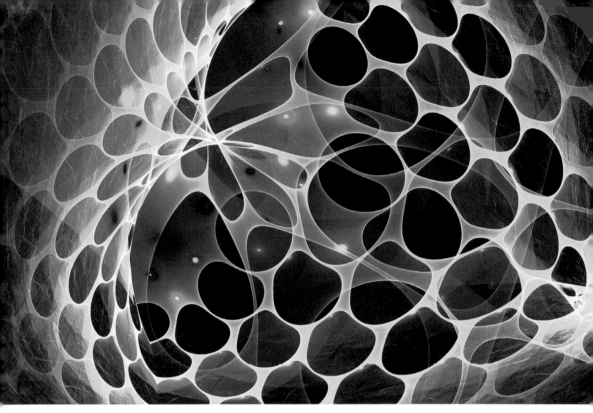

NOVEMBER 21

"If the human mind can understand the universe, it means the human mind is fundamentally of the same order as the divine mind. If the human mind is of the same order as the divine mind, then everything that appeared rational to God as he constructed the universe, its 'geometry,' can also be made to appear rational to the human understanding, and so if we search and think hard enough, we can find a rational explanation and underpinning for everything. This is the fundamental proposition of science."

—ROBERT ZUBRIN, *THE CASE FOR MARS*, 2011

NOVEMBER 22

"He could find how numbers behaved, but he could not explain why.
It was his pleasure to hack his way through the arithmetical jungle, and
sometimes he discovered wonders that more skillful explorers had missed."

—ARTHUR C. CLARKE, *THE CITY AND THE STARS*, 1956

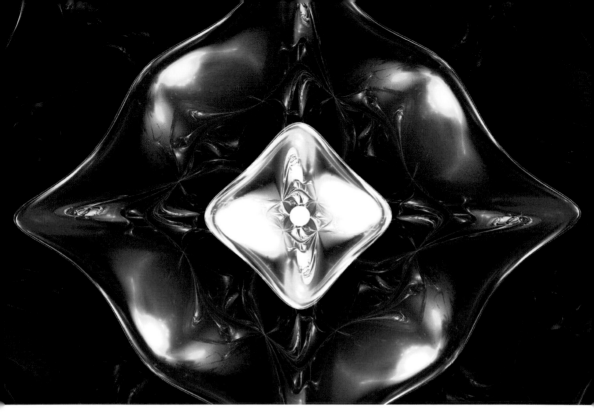

NOVEMBER 23

BORN ON THIS DAY: John Wallis, 1616

"The study of the infinite is much more than a dry, academic game. The intellectual pursuit of the Absolute Infinite is a form of the soul's quest for God. Whether or not the goal is ever reached, an awareness of the process brings enlightenment."

—RUDY RUCKER, *INFINITY AND THE MIND,* 1982

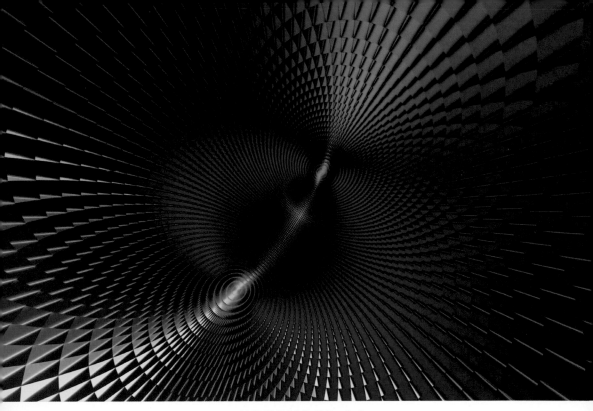

NOVEMBER 24

"I take the positivist viewpoint that a physical theory is just a mathematical model and that it is meaningless to ask whether it corresponds to reality. All that one can ask is that its predictions should be in agreement with observations. . . . All I'm concerned with is that the theory should predict the results of measurements."

—STEPHEN HAWKING, *THE NATURE OF SPACE AND TIME*, 1996

NOVEMBER 25

"Music and math together satisfied a sort of abstract 'appetite,' a desire that was partly intellectual, partly aesthetic, partly emotional, partly, even, physical."

—EDWARD ROTHSTEIN, *EMBLEMS OF MIND*, 1995

NOVEMBER 26

BORN ON THIS DAY: Norbert Wiener, 1894

"The computer, insofar as it solves the equations of mathematical physics, insofar as it is an instrument of oracularity and purports to tell us today what will happen tomorrow, collapses time by making the future appear now."

−PHILIP J. DAVIS AND REUBEN HERSH, *DESCARTES' DREAM*, 1986

NOVEMBER 27

"The truth is that other systems of geometry are possible, yet after all,
these other systems are not spaces but other methods of space measurements.
There is one space only, though we may conceive of many different
manifolds, which are contrivances or ideal constructions invented
for the purpose of determining space."

—PAUL CARUS, "FOUNDATIONS OF MATHEMATICS," *SCIENCE*, 1903

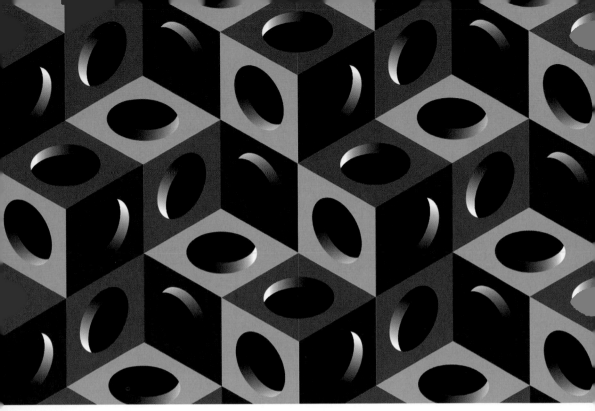

NOVEMBER 28

"It is remarkable that a science which began with the consideration
of games of chance should have become the most important object of human
knowledge. . . . The most important questions of life are indeed,
for the most part, really only problems of probability."

—PIERRE SIMON LAPLACE, *THÉORIE ANALYTIQUE DES PROBABILITÉS*
(ANALYTICAL THEORY OF PROBABILITY), 1812

NOVEMBER 29

"Perfect hexagonal tubes in a packed array. Bees are hard-wired to lay
them down, but how does an insect know enough geometry to lay down a precise
hexagon? It doesn't. It's programmed to chew up wax and spit it out while turning
on its axis, and that generates a circle. Put a bunch of bees on the same surface,
chewing side-by-side, and the circles abut against each other—deform each other into
hexagons, which just happen to be more efficient for close packing anyway."

—PETER WATTS, *BLINDSIGHT*, 2008

NOVEMBER 30

"Computer graphics methods have provided crucial assistance
in many mathematics problems: new minimal surfaces have been found
with the aid of computer graphics, and the visual displays of iterative
maps (in the widely known 'fractal' pictures) make visible patterns that
would never have been noticed by analytic means alone."

—LYNN STEEN, "THE SCIENCE OF PATTERNS," *SCIENCE*, 1988

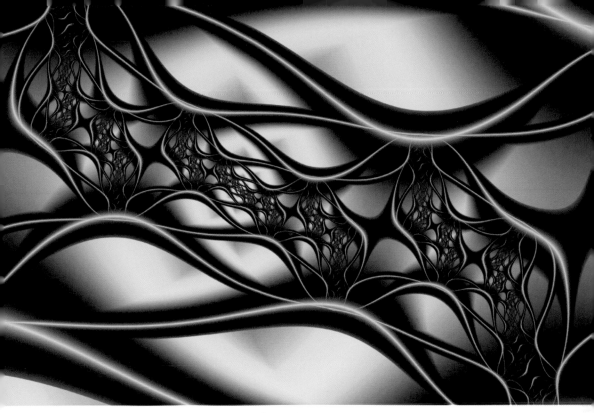

DECEMBER 1

BORN ON THIS DAY: Nikolai Lobachevsky, 1792

"The analytical geometry of Descartes and the calculus of Newton and Leibniz have expanded into the marvelous mathematical method—more daring than anything that the history of philosophy records—of Lobachevsky and Riemann, Gauss and Sylvester. Indeed, mathematics, the indispensable tool of the sciences, defying the senses to follow its splendid flights, is demonstrating today, as it never has been demonstrated before, the supremacy of the pure reason."

—NICHOLAS MURRAY BUTLER, "WHAT KNOWLEDGE IS OF MOST WORTH?," 1895

DECEMBER 2

"Cocoa-buttered girls were stretched out on the public beach in apparently random alignments, but maybe if a weather satellite zoomed in on one of those bodies and then zoomed back out, the photos would show the curving beach itself was another woman, a fractal image made up of the particulate sunbathers. All the beaches pressed together might form female landmasses, female continents, female planets and galaxies. No wonder men felt tense."

—BONNIE JO CAMPBELL, *AMERICAN SALVAGE,* **2009**

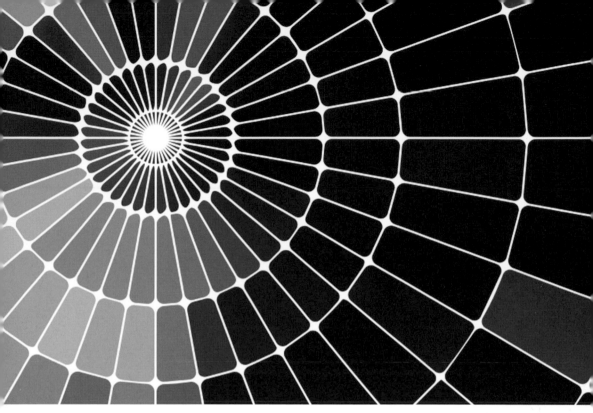

DECEMBER 3

"Geometry was the first exciting course I remember. Instead of
memorizing facts, we were asked to think in clear, logical steps. Beginning
from a few intuitive postulates, far reaching consequences could be derived,
and I took immediately to the sport of proving theorems."

—STEVEN CHU, *NOBEL LECTURES: PHYSICS 1996–2000*, 2002

DECEMBER 4

"One day, as I was walking on the cliff, the idea came to me,
again with the same characteristics of conciseness, suddenness, and
immediate certainty, that arithmetical transformations of indefinite ternary
quadratic forms are identical with those of non-Euclidian geometry."

—JULES HENRI POINCARÉ, *SCIENCE AND METHOD*, 1908

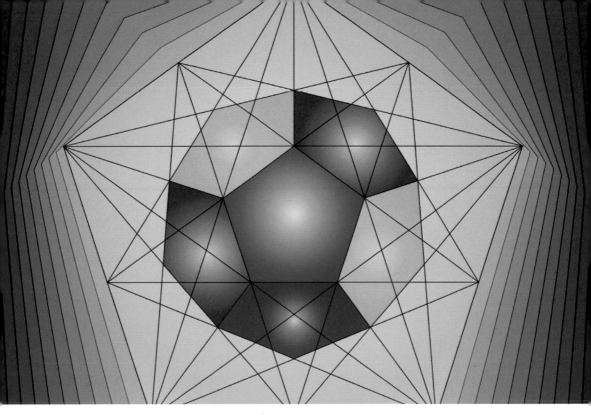

DECEMBER 5

"'Don't you know anything at all about numbers?' 'Well, I don't
think they're very important,' snapped Milo. . . .'NOT IMPORTANT!' roared
the Dodecahedron, turning red with fury. 'Could you have tea for two without the
two—or three blind mice without the three? Would there be four corners of the
earth if there weren't a four? . . . Why, numbers are the most beautiful and
valuable things in the world. Just follow me and I'll show you.' He
turned on his heel and stalked off into the cave."

—NORTON JUSTER, *THE PHANTOM TOLLBOOTH*, 1988

DECEMBER 6

"Even if the rules of nature are finite, like those of chess, might not
science still prove to be an infinitely rich, rewarding game?"

—JOHN HORGAN, "THE NEW CHALLENGES," *SCIENTIFIC AMERICAN*, 1992

DECEMBER 7

"Since geometry is the right foundation of all painting, I have decided to teach its rudiments and principles to all youngsters eager for art."

—ALBRECHT DÜRER, *THE ART OF MEASUREMENT*, 1525

DECEMBER 8

BORN ON THIS DAY: Jacques Salomon Hadamard, 1865; Julia Robinson, 1919

"For a human, there are gigaplex possible thoughts. (A gigaplex is the number written as one followed by a billion zeros.)"

—RUDY RUCKER, *INFINITY AND THE MIND*, 1982

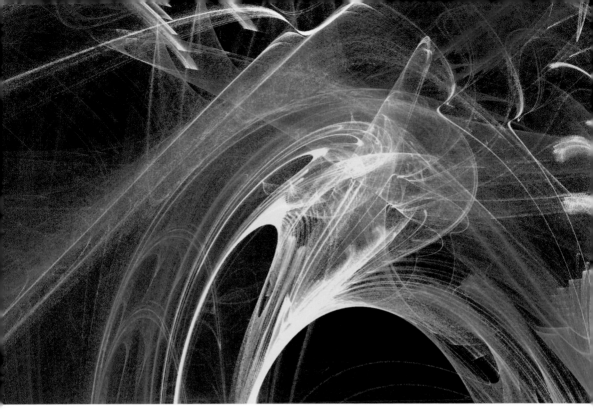

DECEMBER 9

"In geometry I find certain imperfections which I hold to be the reason
why this science, apart from transition into analytics, can as yet make no advance
from that state in which it came to us from Euclid. As belonging to these imperfections,
I consider the obscurity in the fundamental concepts of the geometrical magnitudes
and in the manner and method of representing the measuring of these magnitudes,
and finally the momentous gap in the theory of parallels, to fill which
all efforts of mathematicians have so far been in vain."

—NIKOLAI LOBACHEVSKY, *GEOMETRIC RESEARCHES ON THE THEORY OF PARALLELS*, 1840

DECEMBER 10

BORN ON THIS DAY: Carl Jacobi, 1804

"The ants and their semifluid secretions teach us that pattern, pattern, pattern is the foundational element by which the creatures of the physical world reveal a perfect working model of the divine ideal."

—DON DELILLO, *RATNER'S STAR*, 1976

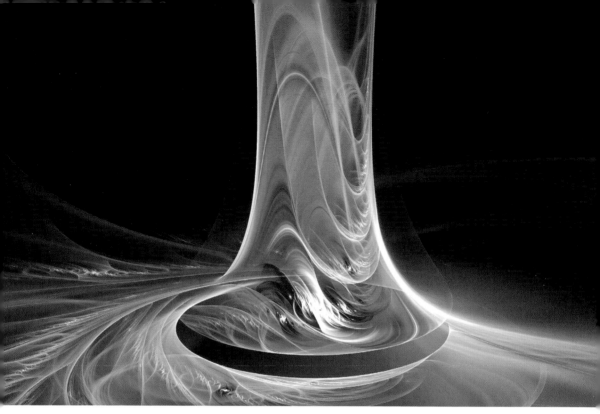

DECEMBER 11

"The object of geometry in all its measuring and computing, is to ascertain with exactness the plan of the great Geometer, to penetrate the veil of material forms, and disclose the thoughts which lie beneath them. When our researches are successful, and when a generous and heaven-eyed inspiration has elevated us above humanity, and raised us triumphantly into the very presence, as it were, of the divine intellect, how instantly and entirely are human pride and vanity repressed, and, by a single glance at the glories of the infinite mind, are we humbled to the dust."

—BENJAMIN PEIRCE, "MATHEMATICAL INVESTIGATION OF THE FRACTIONS WHICH OCCUR IN PHYLLOTAXIS," *INDIANA SCHOOL JOURNAL*, 1856

DECEMBER 18

"By natural selection our mind has adapted itself to the conditions of the external world. It has adopted the geometry most advantageous to the species or, in other words, the most convenient. Geometry is not true, it is advantageous."

—JULES HENRI POINCARÉ, *SCIENCE AND METHOD*, 1908

DECEMBER 19

"For more thousands of years, man has been able to visualize only three dimensions, at right angles to each other. May there not be a fourth dimension, perhaps at right angles to these, in some fashion that we cannot yet picture, or perhaps lying altogether beyond our range of vision? Objects emitting infra-red rays, and lying in such a four-dimensional world, might easily be past our ability to see and our capacity to understand, while existing beside us, nay, in this very hall."

—DONALD WANDREI, *THE BLINDING SHADOWS*, 1934

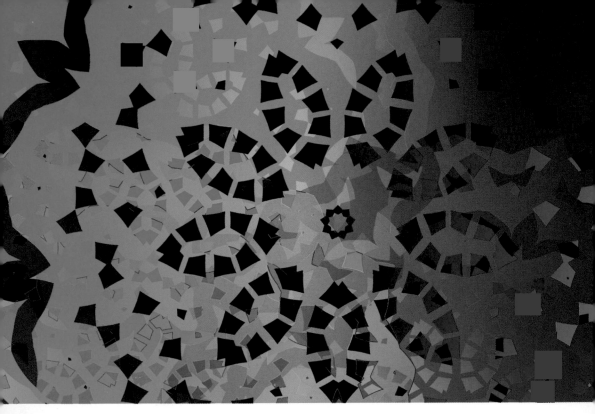

DECEMBER 20

"Coincidences, in general, are great stumbling blocks in the way of
that class of thinkers who have been educated to know nothing of the theory
of probabilities—that theory to which the most glorious objects of human
research are indebted for the most glorious of illustrations."

—EDGAR ALLAN POE, "THE MURDERS IN THE RUE MORGUE," 1841

DECEMBER 21

"If the cosmos were suddenly frozen, and all movement ceased, a survey of its structure would not reveal a random distribution of parts. Simple geometrical patterns, for example, would be found in profusion—from the spirals of galaxies to the hexagonal shapes of snow crystals. Set the clockwork going, and its parts move rhythmically to laws that often can be expressed by equations of surprising simplicity. And there is no logical or a priori reason why these things should be so."

—MARTIN GARDNER, *ORDER AND SURPRISE*, 1950

∞

DECEMBER 22

BORN ON THIS DAY: Srinivasa Ramanujan, 1887

"I remember once going to see [Ramanujan] when he was lying ill at Putney. I had ridden in taxi cab number 1729 and remarked that the number seemed to me rather a dull one, and that I hoped it was not an unfavorable omen. 'No,' he replied, 'it is a very interesting number; it is the smallest number expressible as the sum of two cubes in two different ways.'"

—G. H. HARDY, "THE INDIAN MATHEMATICIAN RAMANUJAN," *THE AMERICAN MATHEMATICAL MONTHLY*, 1937

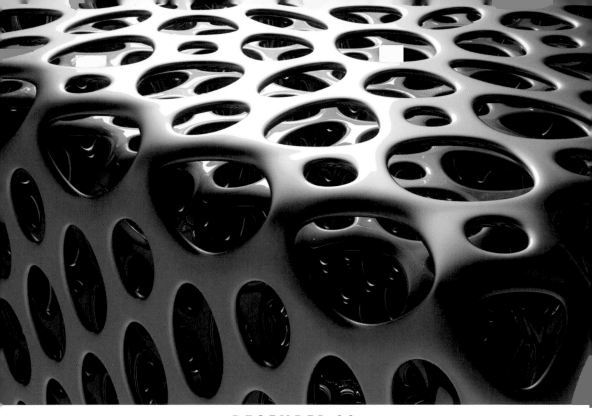

DECEMBER 23

"I regret that it has been necessary for me in this lecture to administer
a large dose of four-dimensional geometry. I do not apologise, because I am
really not responsible for the fact that nature in its most fundamental aspect
is four-dimensional. Things are what they are. . . ."

—ALFRED N. WHITEHEAD, *THE CONCEPT OF NATURE*, 1920

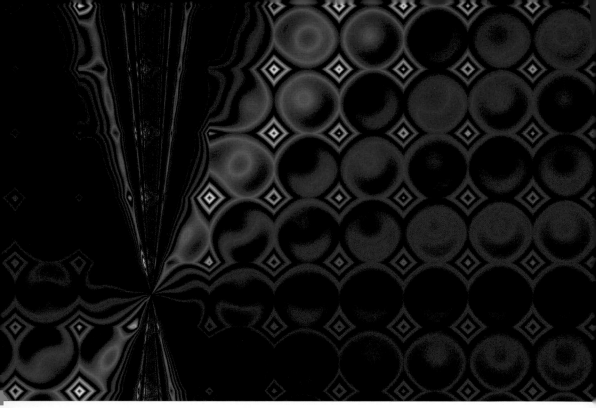

DECEMBER 24

BORN ON THIS DAY: Charles Hermite, 1822

"Wherever there is number, there is beauty."

—PROCLUS, QUOTED IN MORRIS KLINE'S *MATHEMATICAL THOUGHT FROM ANCIENT TO MODERN TIMES*, 1990

DECEMBER 25

BORN ON THIS DAY: Isaac Newton, 1642

"The mathematical take-over of physics has its dangers, as it could tempt us into realms of thought which embody mathematical perfection but might be far removed, or even alien to, physical reality. Even at these dizzying heights we must ponder the same deep questions that troubled both Plato and Immanuel Kant. What is reality? Does it lie in our mind, expressed by mathematical formulae, or is it 'out there'?"

—SIR MICHAEL ATIYAH, "PULLING THE STRINGS," *NATURE*, 2005

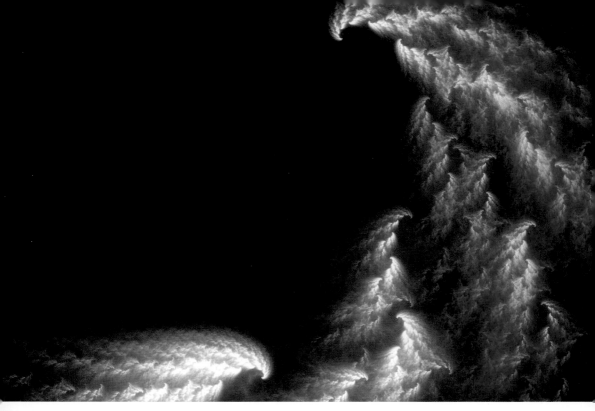

DECEMBER 26

BORN ON THIS DAY: John Horton Conway, 1937

"The essence of life is statistical improbability on a colossal scale."

–RICHARD DAWKINS, *THE BLIND WATCHMAKER*, 1986

364

DECEMBER 27

BORN ON THIS DAY: Jacob Bernoulli, 1654

"We do not live in a time when knowledge can be extended along a pathway smooth and free from obstacles, as at the time of the discovery of the infinitesimal calculus, and in a measure also when in the development of projective geometry obstacles were suddenly removed which, having hemmed progress for a long time, permitted a stream of investigators to pour in upon virgin soil. There is no longer any browsing along the beaten paths; and into the primeval forest only those may venture who are equipped with the sharpest tools."

—HEINRICH BURKHARDT, "MATHEMATISCHES UND WISSENSCHAFTLICHES DENKEN" ("MATHEMATICAL AND SCIENTIFIC THINKING"), C. 1914

DECEMBER 28

BORN ON THIS DAY: John von Neumann, 1903

"Time was when all the parts of the subject were dissevered, when algebra, geometry, and arithmetic either lived apart or kept up cold relations of acquaintance confined to occasional calls upon one another; but that is now at an end; they are drawn together and are constantly becoming more and more intimately related and connected by a thousand fresh ties, and we may confidently look forward to a time when they shall form but one body with one soul."

—JAMES JOSEPH SYLVESTER, PRESIDENTIAL ADDRESS TO SECTION A OF THE BRITISH ASSOCIATION, 1869

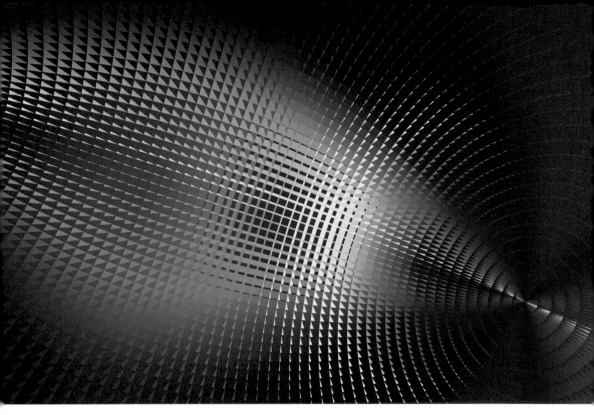

DECEMBER 29

"The theory of limits, on which modern mathematics, and as a corollary,
all modern science and technology depend, is actually a secularized form of the
Scriptural view of reality. Mathematics and Scripture both view the human grasp
of reality as increasingly approximated to, but never identified with,
the human symbols we use to represent reality."

—FORD LEWIS BATTLES, *SOME RUMINATIONS*, 1978

DECEMBER 30

"If geometry were an experimental science, it would not be an exact science, it would be subject to continual revision. . . . In other words the axioms of geometry (I do not speak of those of arithmetic) are only definitions in disguise. What then are we to think of the question: Is Euclidean geometry true? It has no meaning. We might as well ask if the metric system is true and if the old weights and measures are false; if Cartesian coordinates are true and polar coordinates are false. One geometry cannot be more true than another; it can only be more convenient."

—JULES HENRI POINCARÉ, "NON-EUCLIDEAN GEOMETRIES," 1891

DECEMBER 31

BORN ON THIS DAY: Carl Ludwig Siegel, 1896

"One of the endlessly alluring aspects of mathematics is that its thorniest paradoxes have a way of blooming into beautiful theories."

—PHILIP J. DAVIS, "NUMBER," *SCIENTIFIC AMERICAN*, 1964

BERNOULLI, JOHANN (1667–1748) SWITZERLAND

Calculus, L'Hôpital's rule, catenary solution, Bernoulli's rule, Bernoulli's identity, and the "Sophomore's Dream," relating to the formula:

$$\int_0^1 x^{-x}dx = \sum_{n=1}^{\infty} n^{-n} = 1.291285997062663540407282590 5\ldots$$

BIRKHOFF, GEORGE DAVID (1884–1984), USA

Ergodic theorem, number theory, and the chromatic polynomial. Birkhoff proved Poincaré's "Last Geometric Theorem."

BOOLE, GEORGE (1815–1864), ENGLAND

Differential equations, algebraic logic, probability, and Boolean logic (which became the basis of operations for the digital computer).

BOREL, ÉMILE (1871–1956), FRANCE

Measure theory, Borel set, infinite monkey theory, games of strategy, Borel's law of large numbers, Borel summation, and the Borel distribution.

BROUWER, LUITZEN EGBERTUS JAN (1881–1966), HOLLAND

Topology, set theory, measure theory, complex analysis, and intuitionism. Proving the Hairy Ball Theorem. The Brouwer–Hilbert controversy, the Phragmen–Brouwer theorem, and the Brouwer fixed-point theorem.

CANTOR, GEORG (1845–1918) RUSSIA, GERMANY

Set theory and "an infinity of infinites." Transfinite numbers, number theory, the Cantor set, the continuum hypothesis that $C = \aleph_1 = 2^{\aleph_0}$.

CARDANO, GIROLAMO (1501–1576), ITALY

Artis magnae, sive de regulis algebraicis (*Of the Great Art, or The Rules of Algebra*)— also referred to more compactly as *Ars magna*.

Algebra, negative numbers, and solutions to cubic and quartic equations. A fight regarding the solution to $ax^3 + bx + c = 0$.

CARTAN, ÉLIE JOSEPH (1869–1951), FRANCE

Differential geometry, group theory, Lie groups, and spinors (elements of complex vector spaces).

CAUCHY, AUGUSTIN-LOUIS (1789–1857), FRANCE

Cours d'Analyse. Le Calcul infinitesimal (*Summary of Lectures on the Infinitesimal Calculus*). Number theory, calculus, continuity, complex analysis, complex function theory, permutation groups, and abstract algebra. Cauchy's integral theorem: $\oint_c f(z)dz = 0$.

Cauchy's integral formula. The problem of Apollonius, the Cauchy–Binet formula, Cauchy's convergence test, the Cauchy matrix, the Cauchy–Peano theorem, the Cauchy product, Cauchy–Riemann equations, and the Cauchy–Schwarz inequality. Rigorous proof of Taylor's theorem.

CAYLEY, ARTHUR (1821–1895), ENGLAND
"A Memoir on the Theory of Matrices." Cayley–Hamilton theorem, projective geometry, group theory, Cayley numbers, Cayley graph, and Cayley's mousetrap, relating to: 1, 1, 2, 6, 15, 84, 330, 1812, 9978, 65503, . . .

CHEBYSHEV, PAFNUTY (1821–1894), RUSSIA
Probability, statistics, mechanics, and analytical geometry. Bertrand–Chebyshev theorem, Chebyshev polynomials: $T_n(x) = \cos[n \arccos(x)]$.

CHERN, SHIING-SHEN (1911–2004), CHINA, USA
Differential geometry, the Chern–Simons theory, the Chern–Weil theory, the Chern class, integral geometry, value distribution theory of holomorphic functions, minimal submanifolds, and the Chern–Weil homomorphism.

CLAIRAUT, ALEXIS (1713–1765), FRANCE.
Geometry, symmetry, differential equations, Clairaut's theorem (applying to spheroids of revolution), and the formula $g = G[1 + (5m/2 - f) \sin^2\phi]$. Clairaut's relation in differential geometry: $r(t)\cos\theta(t) = $ constant.

CLEBSCH, ALFRED (1833–1872), GERMANY
Algebraic geometry and invariant theory, Clebsch–Gordan coefficients for spherical harmonics, and the Clebsch surface satisfying the two equations $x_0 + x_1 + x_2 + x_3 + x_4 = 0$ and $x_0^3 + x_1^3 + x_2^3 + x_3^3 + x_4^3 = 0$.

CLIFFORD, WILLIAM KINGDON (1845–1879), ENGLAND
Geometric algebra, Clifford algebra (generalizing the real numbers, complex numbers, and several hypercomplex number systems), and Klein–Clifford spaces.

COHEN, PAUL (1934–2007), USA
Neither the continuum hypothesis nor the axiom of choice can be proved from the standard Zermelo–Fraenkel axioms of set theory. The continuum hypothesis is undecidable. Developed the mathematical technique of "forcing."

CONWAY, JOHN HORTON (1937–), ENGLAND
Number theory, theory of finite groups, knot theory, combinatorial game theory, coding theory, the Game of Life, Sprouts, surreal numbers, Conway chained arrow notation,

POINCARÉ, JULES HENRI (1854–1912), FRANCE

Poincaré conjecture (a famous math problem, finally solved around 2003), chaos theory, topology, Poincaré group, the three-body problem, number theory, differential equations, and the special theory of relativity.

POISSON, SIMÉON DENIS (1781–1840), FRANCE

Differential equations, probability, statistics, the Poisson process, the Poisson equation, the Poisson kernel, the Poisson distribution, the Poisson bracket, Poisson algebra, Poisson regression, and the Poisson summation formula.

PÓLYA, GEORGE (1887–1985), HUNGARY

How to Solve It. Probability theory, combinatorics, number theory, numerical analysis, series, Pólya conjecture, Pólya enumeration theorem, Pólya–Vinogradov inequality, Pólya inequality. Pólya distribution, and the Pólya urn model.

PONCELET, JEAN-VICTOR (1788–1867), FRANCE

Traité des propriétés projectives des figures (*Treatise on the Projective Properties of Figures*). Projective geometry, work on Feuerbach's theorem, and the Poncelet–Steiner theorem involving compass and straightedge constructions.

RAMANUJAN, SRINIVASA (1887–1920), INDIA

Number theory, infinite series, continued fractions, mathematical analysis, modular equations, the Landau–Ramanujan constant, mock theta functions, Ramanujan's master theorem, the Ramanujan prime, the Ramanujan–Soldner constant, Ramanujan's sum, Rogers–Ramanujan identities, the Ramanujan–Petersson conjecture, the Ramanujan theta function, nested radicals, and the number 1729 ($1^3 + 12^3 = 9^3 + 10^3$), which is known as the Hardy–Ramanujan number.

RIEMANN, BERNHARD (1826–1866), GERMANY

Analysis, number theory, differential geometry, Riemannian geometry, algebraic geometry, complex manifold theory, the Riemann integral, Riemannian metric, the Riemann hypothesis, and the Riemann zeta function: $\zeta(s) = 1/1^s + 1/2^s + 1/3^s + \ldots$

ROBINSON, JULIA (1919–1985), USA

Diophantine equations and decidability. Well known for her work on decision problems and Hilbert's Tenth Problem.

RUSSELL, BERTRAND (1872–1970), ENGLAND

Principia Mathematica. Logic, set theory, philosophy of mathematics, foundations of mathematics, and Russell's paradox.

IMAGE CREDITS

Juster, Norton. *The Phantom Tollbooth*, New York: Yearling, 1988.

Kasner Edward, and Newman, James. *Mathematics and the Imagination*, New York: Simon & Schuster, 1940.

Kline, Morris. Mathematics: *The Loss of Certainty*, New York: Oxford University Press, 1980.

Lockhart, Paul. *A Mathematician's Lament*, New York: Bellevue Literary Press, 2009.

Manin, Yuri. "Mathematical Knowledge: Internal, Social, and Cultural Aspects," in *Mathematics as Metaphor*, Providence, RI: American Mathematical Society, 2007.

Murakami, Haruki. *1Q84*, New York: Knopf, 2011.

Ouellette, Jennifer. *The Calculus Diaries*, New York, Penguin, 2010.

Penrose, Roger. "What is Reality?" *New Scientist*, vol. 192, no. 2578, pp. 32–39, 2006.

Pickover, Clifford. *A Passion for Mathematics*, Hoboken, NJ: Wiley, 2005.

Poundstone, William. *Labyrinths of Reason*, New York: Anchor, 1988.

Rucker, Rudy. *Infinity and the Mind*, Boston: Birkhäuser, 1982.

Russell, Bertrand. *Mysticism and Logic*, London: George Allen & Unwin, 1918.

Warner, Sylvia. *Mr. Fortune's Maggot*, New York: Viking, 1927.

Weyl, Herman. *The Open World*, New Haven: Yale University Press, 1932.

Whitehead, Alfred North. *An Introduction to Mathematics*, Cambridge: Cambridge University Press, 1911.

Wigner, Eugene. "The Unreasonable Effectiveness of Mathematics in the Natural Sciences," *Communications on Pure and Applied Mathematics*, vol. 13, no. 1, pp. 1–14, 1960.

SELECTED BIBLIOGRAPHY

Atiyah, Michael. "Pulling the Strings," *Nature*, vol. 438, no. 7071, pp. 1081–1082, 2005.

Clarke, Arthur C. *The City and The Stars*, London: Frederick Muller Ltd, 1956.

Clawson, Calvin. *Mathematical Mysteries*, New York: Basic Books, 1996.

Dawkins, Richard. *The Blind Watchmaker*, New York: Norton, 1986.

DeLillo, Don. *Ratner's Star*, New York: Knopf, 1976.

Dyson, Freeman. "Mathematics in the Physical Sciences," *Scientific American*, vol. 211, no. 3, pp. 129–146, 1964.

Einstein, Albert. *The Ultimate Quotable Einstein*, Calaprice, A., ed., Princeton, NJ: Princeton University Press, 2013.

Eves, Howard. *Mathematical Circles*, Boston: Prindle, Weber & Schmidt, 1969.

Farmelo, Graham. *It Must be Beautiful*, London: Granta Books, 2003.

Feynman, Richard. *The Character of Physical Law*, London: BBC, 1965.

Feynman, Richard. *The Feynman Lectures on Physics*, Boston: Addison–Wesley, 1963.

Frenkel, Edward. *Love and Math*, New York: Basic Books, 2013.

Gardner, Martin. "Order and Surprise," *Philosophy of Science*, vol. 17, no. 1, pp. 109–117, 1950.

Giordano, Paolo. *The Solitude of Prime Numbers*, New York: Viking, 2010. (First published in Italian in 2008.)

Gleick, James. *Chaos*, New York: Viking, 1987.

Guillen, Michael. *Five Equations That Changed the World*, New York: Hyperion, 1995.

Hardy, Godfrey Harold. *A Mathematician's Apology*, London: Cambridge University Press, 1941.

Hawking, Stephen. *Black Holes and Baby Universes*, New York: Bantam, 1994.

Hoffman, Paul. *The Man Who Loved Only Numbers*, New York: Hyperion, 1998.

SELBERG, ATLE (1917–2007), NORWAY, USA

Analytic number theory, theory of automorphic forms, spectral theory, the Chowla–Selberg formula, the Selberg sieve, the critical line theorem, Maass–Selberg relations, the Selberg class, Selberg's conjecture, the Selberg integral, the Selberg trace formula, the Selberg zeta function, and the proof of prime number theorem.

SERRE, JEAN-PIERRE (1926–), FRANCE

Algebraic topology, group theory, algebraic geometry, algebraic number theory, the Serre twist sheaf, and Serre fibrations.

SIEGEL, CARL LUDWIG (1896–1981), GERMANY

Number theory, the Thue–Siegel–Roth theorem, the Siegel mass formula, and the Siegel disk.

SMALE, STEPHEN (1930–), USA

Topology, proof of a sphere eversion, dynamical systems, the Smale horseshoe, generalized Poincaré conjecture, handle decomposition, homoclinic orbit, Blum–Shub–Smale machine, regular homotopy, Whitehead torsion, and diffeomorphism.

STEINER, JAKOB (1796–1863), SWITZERLAND

Systematische Entwickelung der Abhängigkeit geometrischer Gestalten von einander (Systematic Development of the Interdependence among Geometrical Configurations). Synthetic geometry, algebraic curves and surfaces, the Steiner tree, and the Steiner chain.

SYLVESTER, JAMES JOSEPH (1814–1897), ENGLAND, USA

Matrix theory, number theory, invariant theory, partition theory, combinatorics, and Sylvester's sequence (each member is the product of the previous members, plus one): 2, 3, 7, 43, 1807, 3263443, 10650056950807, 113423713055421844361000443, . . .

TARSKI, ALFRED (1902–1983), POLAND, USA

Foundations of logic, formal notion of truth, model theory, topology, algebraic logic, metamathematics, abstract algebra, geometry, measure theory, mathematical logic, set theory, and analytic philosophy.

TAYLOR, BROOK (1685–1731), ENGLAND

Taylor's theorem and the Taylor series (which represents a function as an infinite sum of terms that are calculated from the values of the derivatives of a function at a single point):

$$\sum_{n=0}^{\infty} \frac{f^{(n)}(a)}{n!}(x-a)^n$$

Cauchy's integral formula. The problem of Apollonius, the Cauchy–Binet formula, Cauchy's convergence test, the Cauchy matrix, the Cauchy–Peano theorem, the Cauchy product, Cauchy–Riemann equations, and the Cauchy–Schwarz inequality. Rigorous proof of Taylor's theorem.

CAYLEY, ARTHUR (1821–1895), ENGLAND

"A Memoir on the Theory of Matrices." Cayley–Hamilton theorem, projective geometry, group theory, Cayley numbers, Cayley graph, and Cayley's mousetrap, relating to: 1, 1, 2, 6, 15, 84, 330, 1812, 9978, 65503, . . .

CHEBYSHEV, PAFNUTY (1821–1894), RUSSIA

Probability, statistics, mechanics, and analytical geometry. Bertrand–Chebyshev theorem, Chebyshev polynomials: $T_n(x) = \cos[n \arccos(x)]$.

CHERN, SHIING-SHEN (1911–2004), CHINA, USA

Differential geometry, the Chern–Simons theory, the Chern–Weil theory, the Chern class, integral geometry, value distribution theory of holomorphic functions, minimal submanifolds, and the Chern–Weil homomorphism.

CLAIRAUT, ALEXIS (1713–1765), FRANCE.

Geometry, symmetry, differential equations, Clairaut's theorem (applying to spheroids of revolution), and the formula $g = G[1 + (5m/2 - f) \sin^2\phi]$. Clairaut's relation in differential geometry: $r(t)\cos\theta(t) = $ constant.

CLEBSCH, ALFRED (1833–1872), GERMANY

Algebraic geometry and invariant theory, Clebsch–Gordan coefficients for spherical harmonics, and the Clebsch surface satisfying the two equations $x_0 + x_1 + x_2 + x_3 + x_4 = 0$ and $x_0^3 + x_1^3 + x_2^3 + x_3^3 + x_4^3 = 0$.

CLIFFORD, WILLIAM KINGDON (1845–1879), ENGLAND

Geometric algebra, Clifford algebra (generalizing the real numbers, complex numbers, and several hypercomplex number systems), and Klein–Clifford spaces.

COHEN, PAUL (1934–2007), USA

Neither the continuum hypothesis nor the axiom of choice can be proved from the standard Zermelo–Fraenkel axioms of set theory. The continuum hypothesis is undecidable. Developed the mathematical technique of "forcing."

CONWAY, JOHN HORTON (1937–), ENGLAND

Number theory, theory of finite groups, knot theory, combinatorial game theory, coding theory, the Game of Life, Sprouts, surreal numbers, Conway chained arrow notation,

Conway polyhedron notation, classification of finite simple groups, the Angel problem, Conway's soldiers, and the look-and-say sequence: 1, 11, 21, 1211, 111221, 312211, 13112221, 1113213211 . . .

COXETER, HAROLD SCOTT MACDONALD "DONALD" (1907– 2003), ENGLAND, CANADA
Geometry, regular and semiregular polytopes, the Boerdijk–Coxeter helix, the Coxeter functor, the Coxeter group, the Coxeter number, the Coxeter matroid, and Coxeter's loxodromic sequence of tangent circles.

DARBOUX, JEAN-GASTON (1842–1917), FRANCE
Geometry, mathematical analysis, differential equations, differential geometry of surfaces, the Darboux integral, and Darboux's formula for summing infinite series.

DEDEKIND, RICHARD (1831–1916), GERMANY
Abstract algebra (ring theory), foundations of the real numbers, algebraic number theory, the Dedekind cut, and the Dedekind numbers: 2, 3, 6, 20, 168, 7581, 7828354, 2414682040998, 56130437228868755790778 . . .

DESARGUES, GÉRARD (1591–1661), FRANCE
Projective geometry, the Desargues graph, and Desargues's theorem: "Two triangles are in perspective axially if and only if they are in perspective centrally."

DÉSCARTES, RENÉ (1596–1650), FRANCE
La Géométrie (Geometry), Cartesian coordinate system, analytical geometry, philosophy, and Descartes's rule of signs for determining the number of negative or positive real roots of a polynomial.

DIRICHLET, PETER GUSTAV LEJEUNE (1805–1859), GERMANY
Number theory, Fourier series, the Dirichlet function, and Dirichlet's theorem on arithmetic progression.

EISENSTEIN, GOTTHOLD (1823–1852), GERMANY
Number theory, analysis, Eisenstein's reciprocity law, Eisenstein integers, Eisenstein primes, the Eisenstein series, Eisenstein's theorem, and quadratic partitions of prime numbers.

ERDŐS, PAUL (1913–1996), HUNGARY, USA, ISRAEL
Number theory, combinatorics, graph theory, classical analysis, approximation theory, set theory, probability theory, Ramsey theory, proof for the prime number theorem, the Erdős conjecture on arithmetic progressions, the Copeland–Erdős constant (0.2357 1113 1719 2329 3137 4143 . . .) and the Erdös–Borwein constant (1.60669 51524 15291 . . .). He spent most of his life as a vagabond.

EULER, LEONHARD (1707–1783), SWITZERLAND

Calculus, graph theory, power series, Euler's formula [$e^{ix} = \cos(x) + i\sin(x)$], the polyhedron formula ($V - E + F = 2$), Euler's identify $e^{i\pi} + 1 = 0$ (referred to as "the most remarkable formula in mathematics"), and solving the famous Basel problem: $\Sigma 1/n^2 = (1/1^2 + 1/2^2 + 1/3^2 + \ldots + 1/n^2) = \pi/6$. Proved that $2^{31} - 1 = 2,147,483,647$ is a Mersenne prime. Euler's number, $e = 2.7182$ 8182 8459 0452 3536 0287 4713 5266 2497 7572 4709 3699. Euler–Mascheroni constant $\gamma = 0.5772$ 1566 4901 5328 6060 6512 0900 8240 2431 0421 5933 5939.

FATOU, PIERRE (1878–1929), FRANCE

Analysis, holomorphic dynamics, Fatou lemma, Fatou set, and the Fatou–Bieberbach domain.

DE FERMAT, PIERRE (1601–1665), FRANCE

Infinitesimal calculus, number theory, analytic geometry, probability, and Fermat's Last Theorem: $a^n + b^n \neq c^n$ for $n > 2$.

FOURIER, JOSEPH (1768–1830), FRANCE

Fourier series (decomposes periodic functions or signals into the sum of sines and cosines) and the Fourier transform.

FREEDMAN, MICHAEL (1951–), USA

Worked on the Poincaré conjecture in dimension 4.

FREGE, GOTTLOB (1848–1925), GERMANY

Logic, foundations of mathematics, and predicate calculus.

FROBENIUS, FERDINAND GEORG (1849–1917), GERMANY

Elliptic functions, differential equations, group theory, Frobenius–Stickelberger formulae, rational approximations of functions, theory of biquadratic forms, proof for the Cayley–Hamilton theorem, and Frobenius manifolds.

GALOIS, ÉVARISTE (1811–1832), FRANCE

Galois theory, group theory, Galois connections, and the Galois field. He died from wounds inflicted in a duel at the age of twenty.

GARDNER, MARTIN (1914–2010), USA

Gardner brought mathematics to the masses through his popular writing on mathematical games, puzzles, and problems. He wrote on flexagons, Conway's Game of Life, polyominoes, fractals, and various paradoxes.

GAUSS, CARL FRIEDRICH (1777–1855), GERMANY

Disquisitiones Arithmeticae (*Arithmetic Disquisitions*). Number theory, algebra, statistics, analysis, differential geometry, construction of the regular heptadecagon with straightedge and compass, the prime number theorem, and proof of the quadratic reciprocity law.

GÖDEL, KURT (1906–1978), GERMANY, USA

Über formal unentscheidbare Sätze der "Principia Mathematica" und verwandter Systeme ("On Formally Undecidable Propositions of 'Principia Mathematica' and Related Systems"). Incompleteness theorems, Gödel's constructible universe, Gödel's ontological proof of God, and Gödel's slingshot. Starved himself to death.

GOLDBACH, CHRISTIAN (1690–1764), GERMANY

Goldbach's conjecture: every even integer greater than 2 can be expressed as the sum of two primes.

GRASSMANN, HERMANN (1809–1877), GERMANY

Die Lineale Ausdehnungslehre, ein neuer Zweig der Mathematik (*The Theory of Linear Extension, a New Branch of Mathematics*). Linear algebra, exterior algebra, vector spaces, Grassmannians, and the Grassmann number.

GROTHENDIECK, ALEXANDER (1928–), GERMANY, FRANCE

Éléments de géométrie algébrique (*Elements of Algebraic Geometry*) and *Séminaire de Géométrie Algébrique du Bois Marie* (*Algebraic Geometry Seminar of Bois Marie*). Algebraic geometry, commutative algebra, homological algebra, sheaf theory, category theory, topos theory, the first arithmetic Weil cohomology theory, functional analysis, topological tensor products of topological vector space, and Grothendieck's mysterious functor.

HADAMARD, JACQUES SALOMON (1865–1963), FRANCE

Number theory, differential geometry, complex function theory, partial differential equations, the Hadamard product, Hadamard matrices, Hadamard code, and proof of prime number theorem $\pi(x) \sim x/[\ln(x)]$.

HAMILTON, WILLIAM ROWAN (1805–1865), IRELAND

Algebra, Hamiltonians, the Icosian game, and quaternions (extending complex numbers to higher dimensions). The fundamental formula for quaternion multiplication: $i^2 = j^2 = k^2 = ijk = -1$.

HARDY, GODFREY HAROLD (1877–1947), ENGLAND

A Mathematician's Apology. Number theory and mathematical analysis. Mentor of the Indian mathematician Srinivasa Ramanujan. The Hardy–Weinberg principle, Hardy–Ramanujan asymptotic formula, and Hardy–Littlewood conjectures. Hardy once told

Bertrand Russell, "If I could prove by logic that you would die in five minutes, I should be sorry you were going to die, but my sorrow would be very much mitigated by pleasure in the proof."

HAUSDORFF, FELIX (1868–1942), GERMANY

Grundzüge der Mengenlehre (*Basics of Set Theory*). Topology, set theory, measure theory, function theory, the Hausdorff measure, the Hausdorff dimension, and the Hausdorff paradox. In 1942, when he would soon be sent to a Nazi concentration camp, Hausdorff, his wife, and sister-in-law committed suicide.

HERMITE, CHARLES (1822–1901), FRANCE

Number theory, quadratic forms, invariant theory, orthogonal polynomials, elliptic functions, algebra, Hermite polynomials, Hermite interpolation, Hermite normal form, Hermitian operators, and cubic Hermite splines. Proof that e (2.7182 8182 8459 0452...) is transcendental.

HILBERT, DAVID (1862–1943), PRUSSIA, GERMANY

Grundlagen der Geometrie (*Foundations of Geometry*). Invariant theory, axiomatization of geometry, Hilbert spaces, functional analysis, logic, number theory, Hilbert's program, the finiteness theorem, and Hilbert's paradox of the Grand Hotel (a paradox involving infinite sets). In 1900, he presented a famous set of twenty-three math problems that guided a significant amount of mathematics in the twentieth century.

JACOBI, CARL (1804–1851), GERMANY

Fundamenta nova theoriae functionum ellipticarum (*New Foundations of the Theory of Elliptic Functions*). Elliptic functions, differential equations, number theory, and the Jacobian matrix and determinant. First Jewish mathematician to be appointed professor at a German-speaking university (Königsberg University), although he had converted to Christianity by the time of the appointment.

JORDAN, CAMILLE (1838–1921), FRANCE

Cours d'analyse de l'École Polytechnique (*Analysis Course from the École Polytechnique*). Group theory, Jordan curve theorem, Jordan normal form, Jordan measure, the Jordan–Hölder theorem on composition series, Jordan's theorem on finite linear groups, and Galois theory. A Jordan curve is a continuous loop, in the plane, that does not intersect itself.

JULIA, GASTON (1893–1978), FRANCE

Mémoire sur l'itération des fonctions rationnelles (*Notes on the Iteration of Rational Functions*). The fractal Julia set later gained great popularity when Benoît Mandelbrot studied it using computer graphics.

KHAYYÁM, OMAR (1048–1123), PERSIA

Treatise on Demonstration of Problems of Algebra, in which he wrote on the triangular array of binomial coefficients that was later known as Pascal's triangle. Theory of parallels, geometric algebra, and probability. Studied equations such as $x^3 + 200x^2 + 2000$ and $(x + y)^2 = x^2 + 2xy + y^2$.

KLEIN, FELIX (1849–1925), GERMANY

Encyclopedia of Mathematical Sciences Including Their Applications. Group theory, complex analysis, non-Euclidean geometry, the Erlangen Program, Klein bottle (a one-sided closed surface), function theory, number theory, abstract algebra, and the Klein quartic surface.

KOLMOGOROV, ANDREY (1903–1987), RUSSIA

Algorithmic information theory, computational complexity, probability theory, topology, intuitionistic logic, turbulence, classical mechanics, stochastic processes, and Kolmogorov's zero-one law.

KOVALEVSKAYA, SOFIA (1850–1891), RUSSIA

Analysis, differential equations, mechanics, and the Cauchy–Kovalevski theorem. First woman appointed to a full professorship in Northern Europe and the first woman in Europe to hold a doctoral degree in mathematics. The Kovalevskaya Top.

KUMMER, ERNST (1810–1893), GERMANY

Applied mathematics, hypergeometric series, Kummer's function, the Kummer ring, and the Kummer sum.

LAGRANGE, JOSEPH-LOUIS (1736–1813) ITALY, FRANCE

Theorie des fonctions analytiques (*The Theory of Analytical Functions*). Number theory, analysis, calculus of variations, differential equations, probability, calculus, group theory, mechanics, analytical geometry, continued fractions, and Euler–Lagrange equations. Proved the four-square theorem: any natural number can be represented as the sum of four integer squares (e.g. $310 = 17^2 + 4^2 + 2^2 + 1^2$). Proved Wilson's theorem that n is a prime if and only if $(n - 1) + 1$ is a multiple of n ($n > 1$).

LAMBERT, JOHANN HEINRICH (1728–1777), SWITZERLAND, PRUSSIA

Hyperbolic functions in geometry, non-Euclidean geometry, and map projections. First to prove that π is irrational.

LANGLANDS, ROBERT PHELAN (1936–), CANADA

Langlands program, Langlands group, the Langlands–Deligne local constant, Galois groups, algebraic number theory, and automorphic forms.

LAPLACE, PIERRE-SIMON (1749–1827), FRANCE
Statistics, probability, Laplace's equation ($\Delta\varphi = 0$), the Laplace transform, and spherical harmonics.

LEBESGUE, HENRI (1875–1941), FRANCE
"Intégrale, longueur, aire" ("Integral, Length, Area"). Theory of integration, Lebesgue integration, and the Lebesgue measure.

LEGENDRE, ADRIEN MARIE (1752–1833), FRANCE
Éléments de géométrie (*Elements of Geometry*). Legendre polynomials, Legendre transformation, elliptic functions, the least-squares method, and the prime number theorem.

LEIBNIZ, GOTTFRIED WILHELM (1646–1716), GERMANY
Calculus, logic, topology, the Leibniz harmonic triangle, the Leibniz formula for determinants, the Leibniz integral rule, and the Leibniz formula for π: $1 - 1/3 + 1/5 - 1/7 + 1/9 - \ldots = \pi/4$.

LIE, SOPHUS (1842–1899), NORWAY
Continuous symmetry, continuous transformation groups (Lie groups), and Lie algebra.

LIOUVILLE, JOSEPH (1809–1882), FRANCE
Number theory, complex analysis, differential geometry, topology, Liouville's theorem, Liouville numbers, Sturm–Liouville theory, Liouville's equation ($\Delta_0 \log f = -Kf^2$), and the Liouville function: $\lambda(n) = (-1)^{\Omega(n)}$.

LITTLEWOOD, JOHN EDENSOR (1885–1977), ENGLAND
Mathematical analysis, dynamical systems, the Hardy–Littlewood conjecture involving prime numbers, the Littlewood polynomial, and differential equations.

LOBACHEVSKY, NIKOLAI (1792–1856), RUSSIA
Hyperbolic geometry (a form of non-Euclidean geometry).

LORENZ, EDWARD NORTON (1917–2008), USA
Chaos theory and Lorenz attractors. Coined the term butterfly effect.

MANDELBROT, BENOÎT B. (1924–2010), POLAND, FRANCE, USA
The Fractal Geometry of Nature. Fractals and the theory of roughness. The Mandelbrot set ($z_{n+1} = z_n^2 + c$), chaos theory, and the Zipf–Mandelbrot law.

MARKOV, ANDREI (1856–1922), RUSSIA
Stochastic processes (e.g., involving the evolution of some random value through time), Markov chains, and Markov processes.

MATIYASEVICH, YURI VLADIMIROVICH (1947–), RUSSIA
Computability theory. Negative solution of Hilbert's tenth problem, which involves polynomials with integer coefficients.

MERSENNE, MARIN (1588–1648), FRANCE
Mersenne prime numbers (of the form $M_n = 2^n - 1$). The first four Mersenne primes are $M_2 = 3$, $M_3 = 7$, $M_5 = 31$, and $M_7 = 127$.

MILNOR, JOHN WILLARD (1931–), USA
Differential topology, K-theory, dynamical systems, exotic spheres, the Fary–Milnor theorem, Milnor–Thurston kneading theory, and proof of the existence of 7-dimensional spheres with nonstandard differential structure.

MINKOWSKI, HERMANN (1864–1909), LITHUANIA, GERMANY
Geometry of numbers, number theory, theory of relativity, Minkowski space, the Minkowski diagram, and the Minkowski fractal question-mark function (or the "slippery devil's" fractal staircase).

MÖBIUS, AUGUST FERDINAND (1790–1868) GERMANY.
Geometry, number theory, the Möbius strip (a two-dimensional surface with only one side), Möbius transform, Möbius function, Möbius inversion formula, and Möbius transformations: $f(z) = (az + b)/(cz + d)$.

DE MOIVRE, ABRAHAM (1667–1754), FRANCE
The normal distribution, probability theory, and de Moivre's formula, which links complex numbers and trigonometry: $(\cos x + i \sin x)^n = \cos(nx) + i \sin(nx)$.

MONGE, GASPARD (1746–1818), FRANCE
Descriptive geometry, differential geometry, Monge arrays, the Monge–Ampère equation, Monge's theorem involving three circles, and the Monge cone.

NASH, JOHN FORBES, JR. (1928–), USA
Game theory, differential geometry, partial differential equations, Nash equilibrium, Nash embedding theorem, algebraic geometry, and the board game of Hex. Diagnosed with paranoid schizophrenia. Nash was the subject of a 2001 Hollywood movie called *A Beautiful Mind*.

VON NEUMANN, JOHN (1903–1957), HUNGARY, USA
Numerical analysis, functional analysis, set theory, ergodic theory, operator theory, lattice theory, measure theory, geometry, topology, game theory, linear programming, self-replicating machines, statistics, cellular automata, von Neumann paradox, and quantum logic.

NEWTON, ISAAC (1642–1727), ENGLAND

Philosophiæ Naturalis Principia Mathematica ("Mathematical Principles of Natural Philosophy"). Calculus, power series, the binomial theorem with noninteger exponents, optics, classical mechanics, law of universal gravitation, and Newton's method for approximating the roots of a function: $x_{n+1} = x_n - f(x_n)/f'(x_n)$.

NOETHER, AMALIE EMMY (1882–1935), GERMANY

Idealtheorie in Ringbereichen (*Theory of Ideals in Ring Domains*). Abstract algebra, theoretical physics, theories of rings and fields, abstract algebra, hypercomplex numbers, group theory, algebraic invariant theory, elimination theory, topology, Galois theory, and Noether's theorem. Albert Einstein wrote, "Fräulein Noether was the most significant creative mathematical genius thus far produced . . ." Norbert Wiener wrote, "Miss Noether is . . . the greatest woman mathematician who has ever lived."

PASCAL, BLAISE (1623–1662), FRANCE

Traité du triangle arithmétique (*Treatise on the Arithmetical Triangle*). Projective geometry, probability theory, cycloids, and Pascal's triangle for binomial coefficients:

$$
\begin{array}{c}
1 \\
1\ 1 \\
1\ 2\ 1 \\
1\ 3\ 3\ 1 \\
1\ 4\ 6\ 4\ 1 \\
1\ 5\ 10\ 10\ 5\ 1 \\
1\ 6\ 15\ 20\ 15\ 6\ 1
\end{array}
$$

PEACOCK, GEORGE (1791–1858), ENGLAND

Founded the philological or symbolical school of mathematicians. Algebraic theory.

PEANO, GIUSEPPE (1858–1932), ITALY

Mathematical logic, set theory, and the Peano axioms (involving the natural numbers).

PENROSE, ROGER (1931–), ENGLAND

Geometry, Penrose tiling, Twistor theory, geometry of space-time, cosmic censorship, the Weyl curvature hypothesis, Penrose inequalities, and Penrose stairs.

PERELMAN, GRIGORI (1966–), RUSSIA

Proof of the Poincaré conjecture. Riemannian geometry, geometric topology, proof of Thurston's geometrization conjecture, and the analytical and geometric structure of the Ricci flow.

PLÜCKER, JULIUS (1801–1868), GERMANY

Analytical geometry, projective geometry, the Plücker formula, Plücker coordinates, the Plücker surface, and Plücker's conoid, a surface that is a function of two variables: $x = (2xy)/(x^2 + y^2)$.

POINCARÉ, JULES HENRI (1854–1912), FRANCE

Poincaré conjecture (a famous math problem, finally solved around 2003), chaos theory, topology, Poincaré group, the three-body problem, number theory, differential equations, and the special theory of relativity.

POISSON, SIMÉON DENIS (1781–1840), FRANCE

Differential equations, probability, statistics, the Poisson process, the Poisson equation, the Poisson kernel, the Poisson distribution, the Poisson bracket, Poisson algebra, Poisson regression, and the Poisson summation formula.

PÓLYA, GEORGE (1887–1985), HUNGARY

How to Solve It. Probability theory, combinatorics, number theory, numerical analysis, series, Pólya conjecture, Pólya enumeration theorem, Pólya–Vinogradov inequality, Pólya inequality. Pólya distribution, and the Pólya urn model.

PONCELET, JEAN-VICTOR (1788–1867), FRANCE

Traité des propriétés projectives des figures (*Treatise on the Projective Properties of Figures*). Projective geometry, work on Feuerbach's theorem, and the Poncelet–Steiner theorem involving compass and straightedge constructions.

RAMANUJAN, SRINIVASA (1887–1920), INDIA

Number theory, infinite series, continued fractions, mathematical analysis, modular equations, the Landau–Ramanujan constant, mock theta functions, Ramanujan's master theorem, the Ramanujan prime, the Ramanujan–Soldner constant, Ramanujan's sum, Rogers–Ramanujan identities, the Ramanujan–Petersson conjecture, the Ramanujan theta function, nested radicals, and the number 1729 ($1^3 + 12^3 = 9^3 + 10^3$), which is known as the Hardy–Ramanujan number.

RIEMANN, BERNHARD (1826–1866), GERMANY

Analysis, number theory, differential geometry, Riemannian geometry, algebraic geometry, complex manifold theory, the Riemann integral, Riemannian metric, the Riemann hypothesis, and the Riemann zeta function: $\zeta(s) = 1/1^s + 1/2^s + 1/3^s + \ldots$

ROBINSON, JULIA (1919–1985), USA

Diophantine equations and decidability. Well known for her work on decision problems and Hilbert's Tenth Problem.

RUSSELL, BERTRAND (1872–1970), ENGLAND

Principia Mathematica. Logic, set theory, philosophy of mathematics, foundations of mathematics, and Russell's paradox.